Geochemistry

Editor

Sarika Dwivedi

scitus
academics

Geochemistry

Edited by **Sarika Dwivedi**

ISBN: 978-1-68117-139-5

Library of Congress Control Number: 2015936559

www.scitusacademics.com
Box No. 4766, 616 Corporate Way,
Suite 2, Valley Cottage,
NY 10989

Printed in United States of America on Acid Free Paper ∞

Contents

Preface

Geochemistry is the science that uses the tools and principles of chemistry to explain the mechanisms behind major geological systems such as the Earth's crust and its oceans. The realm of geochemistry extends beyond the Earth, encompassing the entire Solar System and has made important contributions to the understanding of a number of processes including mantle convection, the formation of planets and the origins of granite and basalt. "In a word, a comparative geochemistry ought to be launched, before geochemistry can become geology, and before the mystery of the genesis of our planets and their inorganic matter may be revealed."

Editor

Chapter 1

Merging Metagenomics and Geochemistry Reveals Environmental Controls on Biological Diversity and Evolution

Eric B Alsop[1], Eric S Boyd[2, 3], and Jason Raymond[1]

[1]School of Earth and Space Exploration, Arizona State University, ISTB4, Room 795, 781 E. Terrace Rd, Tempe, AZ 85287, USA

[2]Department of Microbiology and Immunology and the Thermal Biology Institute, Montana State University, 109 Lewis Hall, Bozeman, MT 59717, USA

[3]Wisconsin Astrobiology Research Consortium, University of Wisconsin, Weeks Hall, Madison, WI 53706, USA

ABSTRACT

Background

The metabolic strategies employed by microbes inhabiting natural systems are, in large part, dictated by the physical and geochemical properties of the environment. This study sheds light onto the complex relationship between biology and environmental geochemistry using forty-three metagenomes collected from geochemically diverse and globally distributed natural systems. It is widely hypothesized that many uncommonly measured geochemical parameters affect community dynamics and this study leverages the development and application of multidimensional biogeochemical metrics to study correlations between geochemistry and microbial ecology. Analysis techniques such as a Markov cluster-based measure of the evolutionary distance between whole communities and a principal component analysis (PCA) of the geochemical gradients between environments allows for the determination of correlations between microbial community dynamics and environmental geochemistry and provides insight into which geochemical parameters most strongly influence microbial biodiversity.

Results

By progressively building from samples taken along well defined geochemical gradients to samples widely dispersed in geochemical space this study reveals strong links between the extent of taxonomic and functional diversification of resident communities and environmental geochemistry and reveals temperature and pH as the primary factors that have shaped the evolution of these communities. Moreover, the inclusion of extensive geochemical data into analyses reveals new links between geochemical parameters (e.g. oxygen and trace element availability) and the distribution and taxonomic diversification of communities at the functional level. Further, an overall geochemical gradient (from multivariate analyses) between natural systems provides one of the most complete predictions of microbial taxonomic and functional composition.

Conclusions

Clustering based on the frequency in which orthologous proteins occur among metagenomes facilitated accurate prediction of the ordering of community functional composition along geochemical gradients, despite a lack of geochemical input. The consistency in the results obtained from the application of Markov clustering and multivariate methods to distinct natural systems underscore their utility in predicting the functional potential of microbial communities within a natural system based on system geochemistry alone, allowing geochemical measurements to be used to predict purely biological metrics such as microbial community composition and metabolism.

BACKGROUND

The taxonomic and metabolic compositions of microbial communities are both shaped and constrained by the characteristics of their local environment. The characteristics of an environment, in turn, are defined by dynamic physical, geochemical and biological components whose complex interactions are very seldom included in –omics-enabled interrogations of natural communities. This is despite the fact that several recent studies, typically focusing on only a few easily measured environmental parameters, show that natural communities are very tightly tuned—both in overall metabolic function and in community population structure—to nuances of their environment [1-3]. The architecture of natural communities is dictated by competitive and facilitative interactions that function to mold the metabolic strategies responsible for deriving energy and nutrients and maintaining homeostasis against dynamic extracellular environments [4, 5]. These metabolic strategies are encoded within the genomes of individual community members, accessible through advances in sequencing technologies over the past two decades. Although studies comparing community metabolic potential among metagenomes have demonstrated changes in metabolic pathway usage based on environmental geochemistry [6, 7], the focus here is on broad rather than individual metabolic pathway specific deviations in whole community taxonomy and metabolic potential across physical and geochemical gradients.

For a gene to be fixed within a subpopulation of organisms in a complex community, the cognate proteins encoded by the organisms' genomes must function within the geochemical constrains of the environment. The narrow tolerances (e.g. temperature and pH ranges) of some proteins limit the availability of potential habitats for the whole organism, impacting gene flow and, ultimately, colonization ability of the species. For example, the habitat range of photosynthesis along a hydrothermal outflow channel, defined largely by constraints imposed by temperature [8], is a functional limitation that results in a substantial difference in community composition and function, despite negligible differences in physico-chemistry on either side of this upper temperature limit on photosynthesis. Additionally, it is becoming clear that the environmental factors that limit biological function are multidimensional. From the example above, the upper temperature limit for photosynthesis has been discovered to be both pH and sulfide dependent [9-12]. This interdependence between biology and multiple interacting geochemical parameters, as exemplified by the limited distribution of photosynthesis, leads to the hypothesis that there are many additional facets of a community's phenotype that are being shaped by the physical and chemical characteristics of an environment. It stands to reason that many geochemical limitations on a community's phenotype have yet to be discovered—they simply aren't so easy to follow as, for example, the appearance of photosynthetic pigments in a community—yet they may well play central roles in defining community structure and function. The overarching goal of this work is to expand upon current methods of identifying and ultimately quantifying the ecological interactions that most significantly define the structure and function of complex ecosystems.

Here, we integrate sequence data obtained by shotgun community genome sequencing approaches (metagenomics) [13,14] with tools that enable sequence clustering based on a Markov clustering algorithm [15] with BLAST homology [15-17] to categorize metagenomic reads based on evolutionary distance [18-21]. This approach offers a distinct advantage over clustering proteins based on function (i.e. Pfam or KEGG) as the latter approach potentially filters out evolutionary distance information which often extends beyond categories based on protein function [22,23]. Therefore, Markov clustering (and homology-based clustering methods, in general) provide a more direct measure of not only functional differentiation but also overall evolutionary distance among organisms [16]. By applying Markov clustering methods to multiple metagenomic datasets sequence information can be used to determine an overall evolutionary distance between whole communities. The Markov cluster based measure of evolutionary distance can be combined with geochemical analyses allowing statistical techniques including principal components analysis (PCA) and hierarchical clustering to be brought to bear in understanding the interactions between environment and community diversity.

Whole community Markov clustering techniques were first tested using metagenomic datasets gathered along the best available physical, chemical and spatial gradients presently in public databases, and subsequently expanded to include samples gathered from a broader range of environments. This study reveals that several measures of community biodiversity have strong covariance with specific physico-chemical parameters, including temperature, pH, sodium concentration and nitrate availability. A multivariate analysis (PCA) of all geochemical parameters represents clustering by bulk geochemistry and groups metagenomic sites together based on geographic location. Differences in bulk geochemistry covary strongly with community biodiversity, indicating that the composition of the microbial community inhabiting a natural system is determined by a combination of all physical and geochemical parameters of the environment.

RESULT AND DISCUSSION

Validation of Markov Clustering Methods in Met genomic Analysis

Markov clustering methods were initially focused on 22 metagenomic datasets from three studies encompassing very distinct ecosystems, chosen specifically because they extend across steep physical and geochemical gradients: a hydrothermal outflow channel [24], a hypersaline microbial mat [3] and a marine depth profile [2]. These data sets allow for Markov clustering methods to be applied to natural systems with documented community structures, allowing for validation of our methods. The caveat to integrating such a broad range of environmental studies is that, for most met genomic samples, only a few physico-chemical parameters are available (usually temperature and pH). However, our comparisons are bolstered by an inclusion of recently published biogeochemical studies of hydrothermal ecosystems, where metagenome sequencing has been coupled to detailed physical and geochemical analyses [24, 25].

Bison Pool

The Markov clustering approach was first applied to 'Bison Pool', an alkaline hot spring within Yellowstone National Park, USA, where ~500 megabases of Sanger sequencing has been previously compiled from five locations along the outflow channel [24]. Sites 1, 2 and 3 were sampled from the chemotrophic portion of the outflow and sites 4 and 5 were sampled from within the photosynthetic zone. These samples span a 36°C (56.1°C to 92.1°C) temperature gradient, with concomitantly strong changes in a range of geochemical measurements such as dissolved O_2, H_2S, and inorganic nitrogen availability. A dendrogram (Figure 1A) based on Markov cluster analysis of these five metagenomes shows clustering of the photosynthetic sites (4 and 5) separate from the chemotrophic sites (1, 2 and 3), as would be expected based on taxonomic differences among the sites [24]. Additionally, the higher temperature chemotrophic sites cluster separately from site 3, sampled just above the highest temperature where photosynthesis occurs [11] suggesting that this "ecotone" community is transitional between high temperature chemotrophic and lower temperature photosynthetic communities.

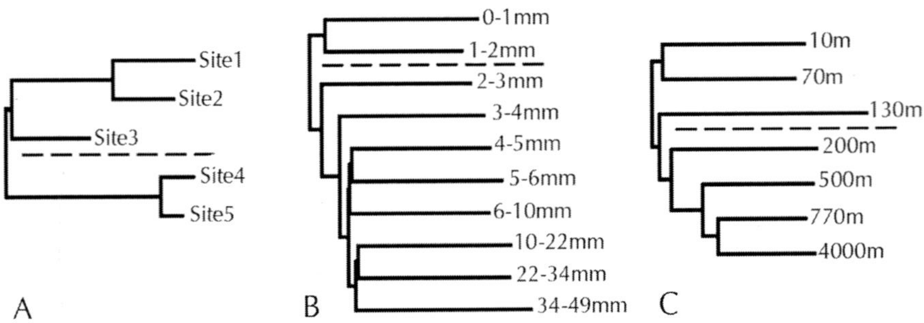

Figure 1: Dendrograms based on Markov cluster dissimilarity in metagenomic datasets obtained for communities inhabiting (A) the outflow chan-

nel of 'Bison Pool' a vertical microbial mat profile from Guerrero Negro, Mexico (B) and an oceanic depth profile from the HOT- ALOHA (C).Dotted lines represent major shifts in microbial community composition: (A) transition between chemotrophic (top) and phototrophic (bottom) metabolisms, (B) transition from an oxic environment near the surface (top) to a hypoxic and sulfidic environment (bottom), and (C)transition from photic zone (top) to aphotic zone (bottom).

Guerrero Negro Hyper saline Microbial Mats

We next applied Markov cluster to a dataset collected from Guerrero Negro, Mexico, which contains approximately 84 me abases of Sanger sequencing of community genomes sampled along millimeter depth scales through ten successive layers of a hyper saline microbial mat [3]. A cluster analysis based dendrogram representing this sample set (Figure 1B) shows clustering of the top 3 mm of the mat separate from the 4 to 50 mm samples with additional clustering of samples from similar depth ranges throughout the mat. As temperatures and pH are not reported as varying over the 49 mm depth profile we must look elsewhere for the cause of the community shifts. Commentary from this study indicates a large drop in oxygen coupled with an increase in H_2S with depth as the driving force for microbial community changes [3]. This transition from an oxic environment to a hypoxic sulfidic environment co-occurs with a major shift in the microbial population and community metabolic strategies, captured in our cluster analysis (Figure 1B). In addition, the clustering of the top 3 mm of mat away from the bottom 46 mm likely correlates with a transition from a mixed phototrophic/chemotrophic community to one supported by chemotrophy and may be related to a shift from aerobic to anaerobic metabolism.

HOT-ALOHA

A marine depth metagenomic profile was also included in this study as the physical and chemical characteristics of the marine water column are known to undergo changes with increasing depth [2]. Samples from the Hawaii Ocean Time-series (HOT) station ALOHA contain approximately 64 megabases of Sanger sequencing of community genomes sampled from seven depths that range from 10 to 4,000 meters [2]. A dendrogram based on Markov clustering (Figure 1C) demonstrates stratification by depth, with nearest neighbors typically coming from similar depths. Clustering occurs with the shallowest samples within the photic zone (10 m and 70 m), representing the separation of samples dominated by photosynthetic metabolisms. Principal component analysis (PCA) of reported geochemical measurements demonstrates that the data can be reduced to two principal components (PC1 and PC2) with combined Eigenvalues explaining 96.8% of the variation (PC1 alone accounts for 85.3% of the variation). A biplot of the two principal components (PC1 verses PC2) (Figure 2) shows separation of tightly clustered photic zone depths (blue points) away from deep water depths (red points). Microbial community changes are reported along the depth gradient with surface waters including Cyanobacteria, Verrucomicrobia, Bacteroidetes and Proteobacteria while deeper waters include members of the Deferribacteres, Planctomycetes, Acidobacteria, Nitrospirae and Proteobacteria phyla (2), despite the depth vector on the biplot being orthogonal to PC1. Along PC1 temperature and dissolved organic carbon (DOC) covary and both exhibit anti-covariation with dissolved inorganic carbon (DIC), nitrite + nitrate (N + N) and dissolved organic phosphorus (DOP). On the whole, PCA demonstrates that depth, as a major component of PC2, is not a good indicator of bulk geochemistry or of community structure or function in marine samples, despite samples clearly segregating into photic and deep water clusters. PCA also suggests other unmeasured variables (most obviously, photon availability) could be driving microbial community changes and underscores an important message: missing or unmeasured physico-chemical variables directly constrain the ability to make meaningful inferences about the interaction between life and environment.

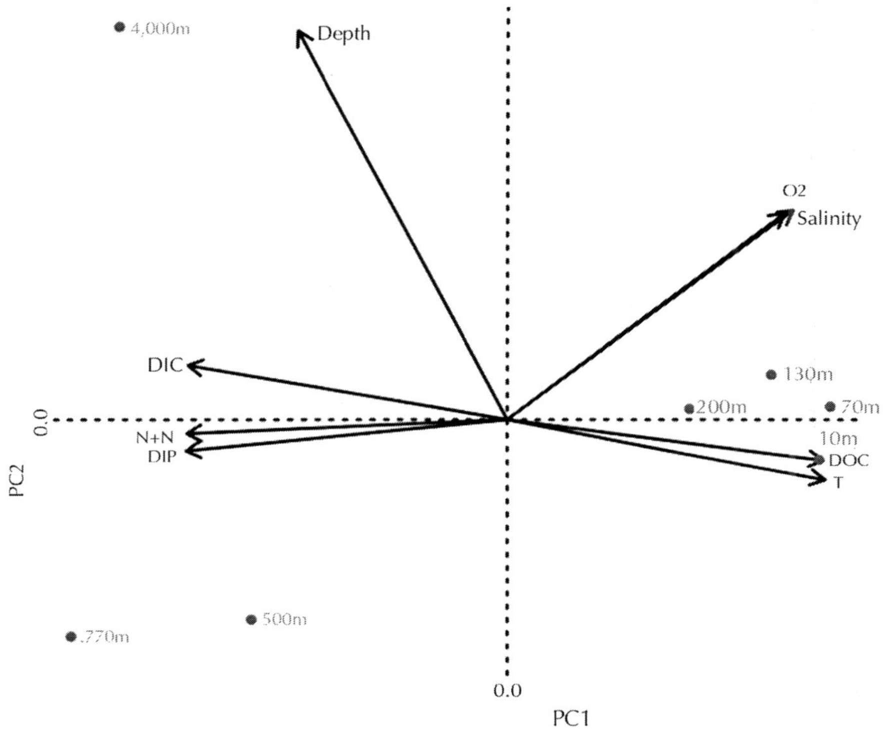

Figure 2: Biplot of the two principal components (PC1 verses PC2) derived from the HOT-ALOHA geochemical data. Sites are colored as chemotrophic (red) and phototrophic (blue).

Expanded Application of Markov Clustering to Diverse Community Met genomes

Role of Environmental Variation in Defining Community Function

The twenty-two metagenomic samples described above occur along diverse spatial and geochemical gradients and, as whole, present key opportunities to connect geochemistry to changes in biological diversity, community structure, and function. Importantly, many additional metagenomes are available that, although not purposefully sampled along continuous gradients, are useful where physico-chemical measurements were made in tandem with biological sampling. Due to the increase in available metagenomic data sets it becomes both statistically feasible and potentially very informative to correlate physical and geochemical differences to changes in the taxonomic and functional diversity of microbial communities. Correlations between geochemistry and biodiversity help identify the key geochemical parameters which shape and constrain taxonomic and functional biodiversity. Figure 3 shows a dendrogram derived from combining the Bison Pool [24], Guerrero Negro [3] and HOT-ALOHA [2] datasets with a hydrothermal sediment metagenome from Great Boiling Spring (GBS), Nevada, USA [26] and twenty additional metagenomes from Yellowstone National Park (YNP) [27-29,25]. Markov cluster analysis of this forty-three metagenome dataset shows a clear separation of hydrothermal and mesothermal sample sites, most notably the separation of the mesothermal Guerrero Negro and HOT-ALOHA sites from the hydrothermal YNP and GBS sites. Note also the temperature segregation within the hydrothermal samples: the lower temperature (phototrophic) YNP sites, including White Creek, Chocolate Pots, and Bison Pool sites 4 and 5 all cluster closest to the mesophilic sites, although the high temperature (chemotrophic) YNP and GBS sites cluster separately. A temperature dependent pattern of clustering due to functional variation between sites is intriguingly similar to the temperature dependent photosynthetic fringe mentioned previously. The successful clustering of whole microbial communities based on temperature differences provides an additional line of evidence supporting the utility of Markov clustering based approaches in comparative genomics analysis.

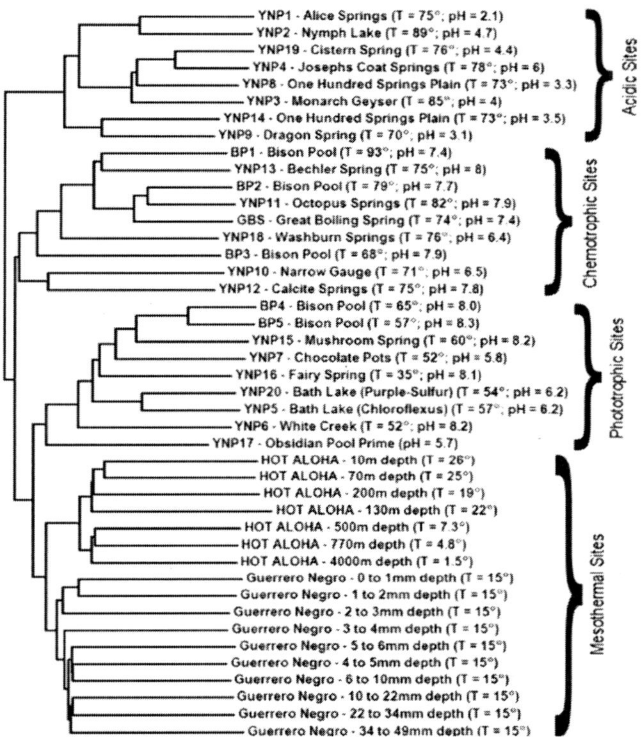

Figure 3: Dendrogram generated from a matrix describing the dissimilarity in Markov clusters associated with forty-three met genomes.

Additionally, a broad level of community segregation based on pH is evident across both GBS and YNP hydrothermal sites, with alkaline samples from Calcite Spring, Washburn Spring, Great Boiling Spring and Bison Pool clustering separately from acidic sites, including Alice Spring, Monarch Geyser and Cistern Spring. A pattern of clustering based on metabolic potential as a function of pH is consistent with previous studies conducted across spatial geochemical gradients in YNP which suggest that pH is the dominant factor shaping the diversification of bacteria and/or archaea at a taxonomic level [30]. The strong influence of pH on the taxonomic and functional composition of hydrothermal communities may reflect different adaptations to deal with acidity [30,31] or may reflect pH-dependent shifts in the energetics associated with inorganic redox couples thought to be fueling these communities [32].

The clustering of communities based predominantly on pH and temperature observed throughout Figure 3 is particularly notable in that it dominates clustering based on biological features, such as the taxonomic or metabolic compositions of communities [6,7]. For instance, the HOT-ALOHA, Guerrero Negro, and YNP datasets all include metagenomes dominated by cyanobacteria whose metabolism is driven by oxygenic photosynthesis, yet clustering of these photosynthetic communities by inorganic factors suggests they have evolved on trajectories optimizing their genomes for conditions specific to each of these environments.

Markov cluster-based evolutionary distances were plotted against temperature (Figure 4A) and pH (Figure 4B) for all pairwise comparisons among the forty-three metagenomes included in this study. Mantel tests [33] show temperature correlates with Markov distance with a Mantel r value of 0.54 ($p<0.001$) and pH correlates with a Mantel r value of 0.41 ($p<0.001$). Although both results show high correlations when compared to other ecological studies using Mantel r values [34-36], it is important to note that the relationships shown in both plots are clearly nonlinear. This nonlinear relationship suggests "envelopes" of allowable space demonstrating that large temperature and/or pH differences drive concomitantly large evolutionary divergences and that communities inhabiting similar temperature and pH ranges are not necessarily evolutionarily related. Certainly organisms and communities have adapted to physico-chemical extremes many times throughout the history of life, and finding unrelated communities occupying similar temperature and pH ranges supports the notion that those adaptations often occur through novel, independent, evolutionary strategies and are not simply the result of adaptation to a small set of environmental variables.

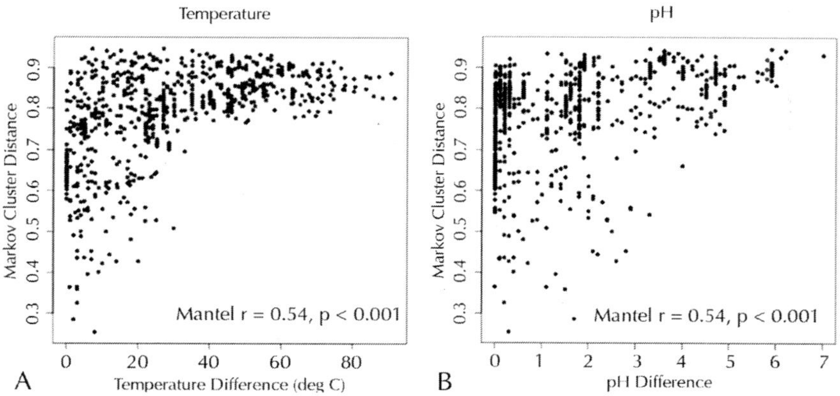

Figure 4: Plots of Markov cluster distance for forty-three metagenomes as a function of temperature (A) and pH (B).

Correlations between microbial community evolutionary divergence and temperature and pH invited deeper exploration of the extensive physical and geochemical data available for some of these metagenomes, in particular a subset of twenty-two metagenomes sequenced as part of several studies of YNP hydrothermal ecosystems [24,27-29,25]. Physical and geochemical metadata includes measurements of temperature, pH, sodium, potassium, calcium, aluminum, iron, magnesium, chloride, phosphorus, silicon, boron, arsenic, zinc, manganese, dissolved oxygen, sulfate, nitrate, sulfide, dissolved organic carbon (DOC) and dissolved inorganic carbon (DIC).

To test for correlation between evolutionary distances and geochemistry, Mantel tests [33] were performed between all geochemical parameters and Markov cluster-based evolutionary distances Several parameters showed slight to moderate correlations [34-36] with evolutionary distances, including chloride (Mantel $r=0.198$, $p=0.007$), zinc (Mantel $r=0.199$, $p=0.010$), DIC (Mantel $r=0.201$, $p=0.018$) and silicon (Mantel $r=0.118$, $p=0.045$). Notably, the parameters showing strongest correlation were, once again, temperature (Mantel $r=0.376$, $p=0.001$) and pH (Mantel $r=0.484$, $p=0.001$). These results reiterate the strong influences temperature and pH have on microbial community evolutionary distance as compared to other geochemical parameters. Importantly, the lack of correlation of Markov cluster distance with some geochemical analytes does not imply lack of a relationship; because these analyses cluster entire metagenomes, the influence of physico-chemistry on individual enzymes and pathways—many of which are known to be strongly dependent on environmental conditions—is, in effect, averaged out.

A covariance matrix based on the twenty included geochemical parameters was used as the basis for a principal component analysis (PCA) of site geochemistry with the Eigenvalues for the first three principal components (PC1, PC2 and PC3) accounting for 61% of the geochemical variation among the twenty-two YNP sites. An overall geochemical distance between YNP sites was calculated by determining the Euclidean distance between YNP sites in (PC1, PC2, PC3) space. A Mantel test was then performed between the overall geochemical distance and the Markov cluster based evolutionary distance for all YNP sites finding a Mantel r value of 0.3861 ($p<0.001$). Although this correlation is weaker than temperature or pH when analyzed individually, a plot of overall geochemical distance verses Markov cluster distance does not display the "envelope" seen in temperature and pH plots. Unlike the "envelope" seen with temperature and pH the overall geochemistry plot is void of points at high community evolutionary distance and low geochemical difference (upper left) indicating that substantially different microbial communities do not inhabit environments with overall similar geochemistry. PCA demonstrates that when analyzed together many site geochemical parameters act in concert to influence the microbial community populating a natural environment. Additionally, PCA hints that the strong correlation with pH might not be due to the concentration of H^+, but to the effect pH has on the speciation of other compounds and the energetic favorability of using these compounds in microbial metabolisms [32].

Role of Geochemical Variation in Defining Community Biodiversity

Finally, we used multivariate techniques to investigate which geochemical parameters most strongly amplify or constrain microbial community diversity. Because biodiversity can be defined quite differently depending on the context and scientific field [37,38], we chose three distinct measures of biological diversity to measure and correlate with environmental metadata. These diversity measurements include: taxonomic diversity (derived from genera counts within each metagenome), functional diversity (derived from metabolic enzyme category (EC) counts within each metagenome), and community complexity (derived from Markov cluster counts within each metagenome). These three measures of diversity were correlated with the twenty geochemical parameters described above (Table 1). Covariance (r from 0.5 to 1) is shaded black and anti-covariance (r from −1 to −0.5) is shaded grey. This analysis shows temperature anti-correlating with genera counts (r = −0.59) and EC counts (r = −0.57) while pH correlates with genera counts (r = 0.71), EC counts (r = 0.62) and Markov cluster counts (r = 0.63). The covariance matrix suggests that low temperature alkaline environments promote community biodiversity whereas high temperature acidic environments constrain biodiversity. Additionally, Markov cluster count is correlating with sodium (r = 0.50) and nitrate (r = 0.50) concentrations while genera count is anti-correlating with zinc concentration (r = −0.54); the functional significance of these relationships is not clear although the strong correlation with sodium may corroborate previous studies suggesting salinity (represented by sodium) as a predominant driver of taxonomic biodiversity [39]. Importantly, the three measurements of biodiversity correlate strongly with one another (bottom-right corner of Table 1). As genetic, functional, and taxonomic diversity are all ultimately encoded at the genetic level and subject to Darwinian evolution, this strong correlation is not surprising, but serves as reassurance that our independently derived measures of biodiversity are in-fact related.

Table 1: Covariance matrix for twenty geochemical variables plus three diversity metrics across twenty two metagenomic sample sites within Yellowstone National Park

	T	pH	Na	K	Ca	Al	Fe	Mg	Cl	NH^4	SO^4	NO^3	P	Si	B	As	Zn	Mn	S^{2-}	O$_2$	Genera	EC	Clusters
T	1.00	−0.25	0.16	0.10	−0.26	0.17	0.02	−0.23	0.25	0.16	0.08	0.27	−0.32	0.36	0.22	0.14	0.24	0.12	−0.06	−0.41	−0.59	−0.57	−0.37
pH	−0.25	1.00	0.36	−0.10	−0.02	−0.61	−0.60	−0.05	−0.26	−0.01	−0.25	0.35	0.01	−0.21	−0.02	−0.03	−0.62	−0.33	0.05	0.48	0.71	0.62	0.63
Na	0.16	0.36	1.00	−0.06	−0.45	−0.31	−0.42	−0.49	0.65	−0.35	−0.16	0.47	−0.48	0.65	0.12	0.19	−0.05	−0.41	−0.51	0.18	0.18	0.35	0.50
K	0.10	−0.10	−0.06	1.00	0.30	−0.14	−0.12	0.80	0.28	0.06	0.16	−0.34	0.00	−0.19	0.75	0.55	0.04	−0.14	0.42	−0.26	−0.41	−0.29	−0.36
Ca	−0.26	−0.02	−0.45	0.30	1.00	−0.12	−0.10	0.99	−0.19	−0.79	0.40	−0.29	0.55	−0.64	−0.10	0.00	−0.14	−0.14	0.62	−0.25	0.37	−0.12	−0.16
Al	0.17	−0.61	−0.31	−0.14	−0.12	1.00	0.93	−0.06	−0.21	−0.09	0.26	−0.17	−0.15	0.22	−0.13	−0.16	0.38	0.19	−0.21	−0.10	−0.38	−0.27	−0.37
Fe	0.02	−0.60	−0.42	−0.12	−0.10	0.93	1.00	−0.05	−0.32	−0.08	0.25	−0.27	0.05	0.10	−0.13	−0.16	0.33	0.31	−0.20	−0.17	−0.31	−0.21	−0.33
Mg	−0.23	−0.05	−0.49	0.80	0.99	−0.06	−0.05	1.00	−0.23	−0.02	0.49	−0.27	0.52	−0.65	−0.11	−0.03	−0.13	−0.15	0.68	−0.28	0.05	−0.14	−0.18
Cl	0.25	−0.26	0.65	0.28	−0.19	−0.21	−0.32	−0.23	1.00	−0.20	−0.45	0.01	−0.40	0.65	0.25	0.30	0.40	−0.35	−0.27	−0.08	−0.33	−0.14	−0.03
NH$_4$	0.16	−0.01	−0.35	0.06	−0.79	−0.09	−0.08	−0.02	−0.20	1.00	0.77	0.30	0.07	−0.19	0.30	0.20	−0.12	−0.08	0.57	−0.19	−0.01	−0.03	−0.01
SO$_4$	0.08	−0.25	−0.16	0.16	0.40	0.26	0.25	0.49	−0.45	0.77	1.00	0.02	0.27	−0.44	0.09	−0.01	−0.01	−0.07	0.79	−0.36	−0.12	−0.21	−0.27
NO$_3$	0.27	0.35	0.47	−0.34	−0.29	−0.17	−0.27	−0.27	0.01	0.30	0.02	1.00	−0.21	0.29	−0.10	−0.12	−0.36	−0.33	0.00	−0.01	0.36	0.44	0.50
P	−0.32	0.01	−0.48	0.00	0.55	−0.15	0.05	0.52	−0.40	0.07	0.27	−0.21	1.00	−0.60	−0.18	−0.11	−0.20	0.12	0.41	−0.24	0.23	0.36	0.04
Si	0.36	−0.21	0.65	−0.19	−0.64	0.22	0.10	−0.65	0.65	−0.19	−0.44	0.29	−0.60	1.00	0.12	0.16	0.33	−0.23	−0.59	0.15	−0.20	0.04	0.17
B	0.22	−0.02	0.12	0.75	−0.10	−0.13	−0.13	−0.11	0.25	0.30	0.09	−0.10	−0.18	0.12	1.00	0.89	−0.06	−0.35	0.15	−0.22	−0.39	−0.21	−0.21
As	0.14	−0.03	0.19	0.55	0.00	−0.16	−0.15	−0.03	0.30	0.20	−0.01	−0.12	−0.11	0.16	0.89	1.00	−0.11	−0.06	−0.02	−0.14	−0.29	−0.15	−0.07
Zn	0.24	−0.62	−0.05	0.04	−0.14	0.38	0.33	−0.13	0.40	−0.12	−0.01	−0.36	−0.20	0.33	−0.06	−0.11	1.00	0.07	−0.16	−0.20	−0.54	−0.38	−0.36
Mn	0.12	−0.33	−0.41	−0.14	−0.14	0.19	0.31	−0.15	−0.35	−0.08	−0.07	−0.33	0.12	−0.23	−0.35	−0.06	0.07	1.00	−0.23	−0.16	−0.27	−0.35	−0.34
S^{2-}	−0.06	0.05	−0.51	0.42	0.62	−0.21	−0.20	0.68	−0.27	0.57	0.79	0.00	0.41	−0.59	0.15	−0.02	−0.16	−0.23	1.00	−0.35	0.09	−0.04	−0.10
O$_2$	−0.41	0.48	0.18	−0.26	−0.25	−0.10	−0.17	−0.28	−0.08	−0.19	−0.36	−0.01	−0.24	0.15	−0.22	−0.14	−0.20	−0.16	−0.35	1.00	0.46	0.48	0.42
Genera	−0.59	0.71	0.18	−0.41	0.37	−0.38	−0.31	0.05	−0.33	−0.01	−0.12	0.36	0.23	−0.20	−0.39	−0.28	−0.54	−0.27	0.09	0.46	1.00	0.91	0.87
EC	−0.57	0.62	0.35	−0.29	−0.12	−0.27	−0.21	−0.14	−0.14	−0.03	−0.21	0.44	0.36	0.04	−0.21	−0.15	−0.38	−0.35	−0.04	0.48	0.91	1.00	0.92
Clusters	−0.37	0.63	0.50	−0.36	−0.16	−0.37	−0.33	−0.18	−0.03	−0.01	−0.27	0.50	0.04	0.17	−0.21	−0.07	−0.36	−0.34	−0.10	0.42	0.87	0.92	1.00

Covariance between 0.5 and 1 are shaded black and anti-covariance between −1 and −0.5 are shaded grey.

Alsop *et al.*

Alsop *et al.* *BMC Ecology* 2014 14:16, doi: 10.1186/1472-6785-14-16

A biplot (Figure 5) generated from a PCA of the geochemical and diversity correlation matrix shows where the metagenomic sites lie within physico-chemical and biodiversity space. Archaeal-dominated sites (YNP 1, 2, 3, 4, 8, 14 and 19) populate the upper right quadrant of the biplot, hinting that the geochemical parameters associated with these sites exclude bacterial life that lack functional adaptations to inhabit these springs [40]. The photosynthetic mat samples collected from the Lower Geyser Basin (BP 4, 5, YNP 6, 15 and 16) show clustering, but separate from the photosynthetic mats found at Mammoth hot springs (YNP 5 and 20). Aquificales-dominated sites (YNP 10, 11, 12 and 13) do not cluster with each other, but instead cluster based on geographic proximity and geochemical similarity. For instance, YNP 11 (Octopus Spring) clusters with Bison Pool site 1; both springs are alkaline and are proximal geographically. Likewise, YNP 14 (One Hundred Springs Plain) clusters with other Norris Geyser Basin springs such as YNP 3 (Monarch Geyser). Bison Pool (BP) sites illustrate the strengths of PCA for correlating this multidimensional dataset: all five BP sites are in the same region of the biplot due to the overall similar geochemistry among sites and, further, are aligned in a linear fashion parallel to the temperature vector (due to the 32°C temperature gradient along the outflow). The placement of similar sites on the PCA biplot illustrates the predictive power of PCA as implemented here; one could reasonably predict where a new site might plot based on measurements of only a handful of well-chosen biological and physico-chemical parameters.

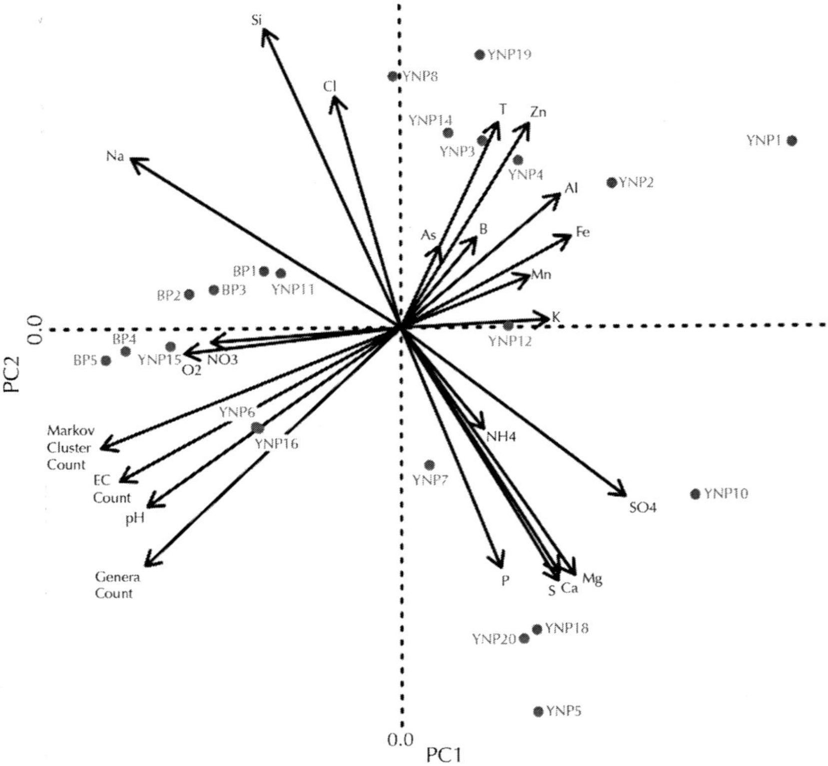

Figure 5: Biplot generated from a principal component analysis (PCA) of the twenty-two YNP sites with individual sample sites and diversity metrics depicted. Sites are colored as chemotrophic (red) and phototrophic (blue).

CONCLUSIONS

Markov cluster based comparisons of metagenomes coupled with multivariate analyses identified many key physical and geochemical parameters which are responsible for shaping microbial community composition, function, and complexity. Most metagenomic datasets include very limited (or no) environmental metadata; here we focused on a subset of metagenomes with detailed measurements of pH and temperature, and a subset of these (from hydrothermal systems) with 18 additional geochemical measures that could be compared. Our analysis supports the strong role that pH and temperature play in influencing microbial community composition and function, accounting for the highest average Mantel correlations (0.48 for pH and 0.38 for temperature) to evolutionary distance between metagenomes. Importantly, upon the inclusion of additional geochemical parameters it was found that the availability of carbon compounds as well as micronutrients such as iron and zinc all correlate (or anticorrelate) with diversity measures. Multivariate analyses suggest that these biology-environment interactions are multidimensional: techniques integrating many physical and chemical measurements performed as well as or better than nearly all of the individual parameters at predicting differences in biodiversity. This demonstrates that the parameters typically measured as part of metagenome studies (temperature, pH, depth) can be substantially improved upon in attempts to explain or predict biological variability as a function of environmental dynamics.

Finally, this analysis lays the groundwork for predicting community metabolism and various metrics of diversity based on site geochemistry. For example, PCA analysis of YNP community metagenomes and bulk geochemistry can predict biological properties of an unknown site based on geochemistry, and vice versa. Future metagenomic studies can continue to improve the resolving power of these predictions simply by including a small number of relatively straightforward measurements of physical and geochemical conditions along with biological sampling. This study represents an important advance toward predictive understanding of biology-environment interactions, and a compelling justification for coordinating environmental/geochemical measurements in –omics-enabled studies of natural environments.

METHODS

All metagenomic datasets were downloaded as inferred amino acid sequences from the Joint Genome Institute Integrated Microbial Genomes with Microbiome Samples (JGI IMG/M) [26] web server. All metagenomic datasets were combined into a single FASTA file and compared using a complete all-verses-all NCBI BLAST [41]. BLAST results were then parsed to hits with e-values better than 10^{-40}. Parsed BLAST results were fed into the mcl [16] Markov clustering algorithm using an inflation value of 1.2. The mcl algorithm generates a network where nodes represent individual genes or proteins and the edges between them are weighted based on some measure of homology (here, BLAST e value, though in principle any homology score can be used). The heuristic then performs Markov walks across this network—quasi-random walks between nodes whose probability of traversal depends upon the strength of the edge connecting them (dependent on the homology score). Network edges are strengthened or weakened based on the number of traversals during each iteration, with the inflation parameter influencing how rapidly edges are strengthened and whether or not an edge is 'severed'. This procedure of random walks followed by edge strengthening and/or culling is iterated until convergence, typically when no edges are strengthened or lost from the network. At convergence, nodes which remain connected are output as Markov clusters. BLAST e-value cut-off and MCL inflation values were chosen such as to maximize the inclusion of homologous proteins into resultant Markov clusters [16,24]. Perl scripts were written to determine the Jaccard (binary) dissimilarity between metagenomes by summing the total clusters shared by a pair of metagenomes and dividing by the total number of Markov clusters in each metagenome pair, resulting in a dissimilarity value of 0 (all Markov clusters occur in both metagenomes) and a dissimilarity value of 1 (no Markov clusters occur in both metagenomes). Perl scripts were then used to convert Jaccard dissimilarities among all metagenomes into distance matrices. Distance matrices were converted into dendrograms using the NEIGHBOR program within the PHYLIP software package[42]. Markov cluster distances were calculated as the total branch length distance between dendrogram terminal nodes (leaves) using the TreeIO module within the BioPerl [43] software package.

Perl scripts were used to generate dissimilarity matrices using all geochemical gradients (differences) between sampled locations. Additional, Perl scripts were used to generate dissimilarity matrices from the calculated Markov cluster distances among metagenomes. Mantel tests were performed in R [44] using the Vegan package [45] function "mantel". Mantel tests were completed with the Pearson method using 1,000 permutations. Mantel test results were plotted as geochemical difference verses Markov cluster distance for all metagenome pairs with separate plots for each geochemical parameters.

Correlation matrices and PCA analyses were completed using the base package R (version 2.11.1)[44] with the raw geochemical measurements as input. Correlation matrices were calculated using the "cor" function in R using the Pearson correlation method. PCA was completed using the Vegan package [45] functions "rda" with scaling enabled. PCA results were graphed using the "biplot" function with scaling of species and sites.

All metagenome sequences were compared to the NCBI non-redundant (nr) database and the KEGG[46] database using NCBI BLAST [41]. EC count was determined by tallying all unique EC numbers with a minimum of two hits from the BLAST versus the KEGG database. Genus counts were completed by tallying unique genus level hits from the best BLAST hit to the nr database, if one existed with a e-value better than 10^{-40}. Tally was parsed to include only genera within 80th percentile of total hits, allowing genera with very low counts to be excluded from the analysis. Markov cluster counts were a tally of the number of Markov clusters within a met genome.

Perl scripts developed for use in this study are freely available from the study's coauthors.

AUTHORS' CONTRIBUTIONS

Conceived and designed the experiments: EBA ESB JR. Performed the computations: EBA. Analyzed the data: EBA ESB JR. Wrote the paper: EBA ESB JR. All authors read and approved the final manuscript.

ACKNOWLEDGEMENTS

We thank Everett Shock, Jack Farmer, and Ariel Anbar for extensive discussion and helpful feedback in carrying out this research. This work was funded by a NASA Astrobiology Institute (NAI) "Follow the Elements" grant (JR) and grants NNX08AP61G (JR) and NNX13AI11G (ESB) from the NASA Exobiology and Evolutionary Biology program. The Wisconsin Astrobiology Research Consortium is supported by the NAI (NNA13AA94A) to ESB.

REFERENCES

1. Allen EE, Banfield JF: Community genomics in microbial ecology and evolution. *Nat Rev Microbiol* 2005, 3:489-498.
2. DeLong EF, Preston CM, Mincer T, Rich V, Hallam SJ, Frigaard N-U, Martinez A, Sullivan MB, Edwards R, Brito BR, Chisholm SW, Karl DM: Community genomics among stratified microbial assemblages in the ocean's interior. *Science* 2006, 311:496-503.
3. Kunin V, Raes J, Harris JK, Spear JR, Walker JJ, Ivanova N, von Mering C, Bebout BM, Pace NR, Bork P, Hugenholtz P: Millimeter-scale genetic gradients and community-level molecular convergence in a hypersaline microbial mat. *Mol Syst Biol* 2008, 4:198.
4. Little AEF, Robinson CJ, Peterson SB, Raffa KF, Handelsman J: Rules of engagement: interspecies interactions that regulate microbial communities. *Annu Rev Microbiol* 2008, 62:375-401.
5. Hamilton TL, Koonce E, Howells A, Havig JR, Jewell T, De La Torre JR, Peters JW, Boyd ES:Competition for Ammonia Influences the Structure of Chemotrophic Communities in Geothermal Springs. *Appl Environ Microbiol* 2013, 80:653-661. No. 2
6. Dinsdale EA, Edwards RA, Hall D, Angly F, Breitbart M, Brulc JM, Furlan M, Desnues C, Haynes M, Li L, McDaniel L, Moran MA, Nelson KE, Nilsson C, Olson R, Paul J, Brito BR, Ruan Y, Swan BK, Stevens R, Valentine DL, Thurber RV, Wegley L, White BA, Rohwer F: Functional metagenomic profiling of nine biomes. *Nature* 2008, 452:629-632.
7. Gianoulis TA, Raes J, Patel PV, Bjornson R, Korbel JO, Letunic I, Yamada T, Paccanaro A, Jensen LJ, Snyder M, Bork P, Gerstein MB: Quantifying environmental adaptation of metabolic pathways in metagenomics. *PNAS* 2009, 106:1374-1379. No. 2
8. Brock TD: Micro-organisms adapted to High Temperatures. *Published online* 1967, 214:882-885. doi:101038/214882a0
9. Boyd ES, Hamilton TL, Spear JR, Lavin M, Peters JW: [FeFe]-hydrogenase in Yellowstone National Park: evidence for dispersal limitation and phylogenetic niche conservatism. *ISME J* 2010, 4:1485-1495.

10. Hamilton TL, Lange RK, Boyd ES, Peters JW: Biological nitrogen fixation in acidic high-temperature geothermal springs in Yellowstone National Park, Wyoming. *Environ Microbiol* 2011, 13:2204-2215.

11. Cox A, Shock EL, Havig JR: The transition to microbial photosynthesis in hot spring ecosystems. *Chem Geol* 2011, 280:344-351.

12. Boyd ES, Fecteau KM, Havig JR, Shock EL, Peters JW: Modeling the Habitat Range of Phototrophs in Yellowstone National Park: Toward the Development of a Comprehensive Fitness Landscape. *Front Microbiol* 2012, 3:221.

13. Tyson GW, Chapman J, Hugenholtz P, Allen EE, Ram RJ, Richardson PM, Solovyev VV, Rubin EM, Rokhsar DS, Banfield JF: Community structure and metabolism through reconstruction of microbial genomes from the environment. *Nature* 2004, 428:37-43.

14. Venter JC, Remington K, Heidelberg JF, Halpern AL, Rusch D, Eisen JA, Wu D, Paulsen I, Nelson KE, Nelson W, Fouts DE, Levy S, Knap AH, Lomas MW, Nealson K, White O, Peterson J, Hoffman J, Parsons R, Baden-Tillson H, Pfannkoch C, Rogers Y-H, Smith HO: Environmental Genome Shotgun Sequencing of the Sargasso Sea. *Science* 2004, 304:66-74.

15. Van Dongen SM: *Graph clustering by flow simulation*. 2000.

16. Enright AJ, Van Dongen S, Ouzounis CA: An efficient algorithm for large-scale detection of protein families. *Nucleic Acids Res* 2002, 30:1575-1584.

17. Swingley WD, Blankenship RE, Raymond J: Integrating Markov Clustering and Molecular Phylogenetics to Reconstruct the Cyanobacterial Species Tree from Conserved Protein Families. *Mol Biol Evol* 2008, 25:643-654.

18. Shih Y-K, Parthasarathy S: Identifying functional modules in interaction networks through overlapping Markov clustering. *Bioinformatics* 2012, 28:i473-i479.

19. Tanaseichuk O, Borneman J, Jiang T: Separating metagenomic short reads into genomes via clustering. *Algorithms Mol Biol* 2012, 7:27.

20. Klatt CG, Wood JM, Rusch DB, Bateson MM, Hamamura N, Heidelberg JF, Grossman AR, Bhaya D, Cohan FM, Kühl M, Bryant DA, Ward DM: Community ecology of hot spring cyanobacterial mats: predominant populations and their functional potential. *ISME J* 2011, 5:1262-1278.

21. Yamada T, Waller AS, Raes J, Zelezniak A, Perchat N, Perret A, Salanoubat M, Patil KR, Weissenbach J, Bork P: Prediction and identification of sequences coding for orphan enzymes using genomic and metagenomic neighbours. *Mol Syst Biol* 2012, 8:581.

22. Fukami-Kobayashi K, Tomoda S, Gō M: Evolutionary clustering and functional similarity of RNA-binding proteins. *FEBS Lett* 1993, 335:289-293.

23. Tatusov RL, Koonin EV, Lipman DJ: A Genomic Perspective on Protein Families. *Science* 1997, 278:631-637.

24. Swingley WD, Meyer-Dombard DR, Shock EL, Alsop EB, Falenski HD, Havig JR, Raymond J:Coordinating Environmental Genomics and Geochemistry Reveals Metabolic Transitions in a Hot Spring Ecosystem. *PLoS One* 2012, 7:e38108.

25. Inskeep WP: The YNP Metagenome Project: Environmental Parameters Responsible for Microbial Distribution in the Yellowstone Geothermal Ecosystem. *Front Microbiol* 2013, 4:67.

26. Grigoriev IV, Nordberg H, Shabalov I, Aerts A, Cantor M, Goodstein D, Kuo A, Minovitsky S, Nikitin R, Ohm RA, Otillar R, Poliakov A, Ratnere I, Riley R, Smirnova T, Rokhsar D, Dubchak I: The Genome Portal of the Department of Energy Joint Genome Institute. *Nucl Acids Res* 2011, 40:D25-32.

27. Inskeep WP, Klatt CG: Community Structure and Function of High-temperature Chlorophototrophic Microbial Mats Inhabiting Diverse Geothermal Environments. *Front Microbiol* 2013, 4:106.

28. Inskeep WP: Phylogenetic and functional analysis of metagenome sequence from high-temperature archaeal habitats demonstrate linkages between metabolic potential and geochemistry. *Front Microbiol* 2013, 4:95.

29. Inskeep WP, Takacs Vesbach C: Metagenome Sequence Analysis of Filamentous Microbial Communities Obtained from Geochemically Distinct Geothermal Channels Reveals

Specialization of Three Aquificales Lineages. *Front Microbiol* 2013, 4:84.

30. Boyd ES, Wang J, He L: The role of tetraether lipid composition in the adaptation of thermophilic archaea to acidity. *Front Microbiol* 2013, 4:62.
31. Pearson A, Pi Y, Zhao W, Li W, Li Y, Inskeep W, Perevalova A, Romanek C, Li S, Zhang CL:Factors Controlling the Distribution of Archaeal Tetraethers in Terrestrial Hot Springs. *Appl Environ Microbiol* 2008, 74:3523-3532.
32. Shock EL, Holland M, Meyer-Dombard D, Amend JP, Osburn GR, Fischer TP: Quantifying inorganic sources of geochemical energy in hydrothermal ecosystems, Yellowstone National Park, USA. *Geochim Cosmochim Acta* 2010, 74:4005-4043.
33. Mantel N, Haenszel W: Statistical aspects of the analysis of data from retrospective studies of disease. *J Natl Cancer Inst* 1959, 22:719-748.
34. Turnbaugh PJ, Hamady M, Yatsunenko T, Cantarel BL, Duncan A, Ley RE, Sogin ML, Jones WJ, Roe BA, Affourtit JP, Egholm M, Henrissat B, Heath AC, Knight R, Gordon JI: A core gut microbiome in obese and lean twins. *Nature* 2009, 457:480-484.
35. Toulza E, Tagliabue A, Blain S, Piganeau G: Analysis of the Global Ocean Sampling (GOS) Project for Trends in Iron Uptake by Surface Ocean Microbes. *PLoS One* 2012, 7(2):e30931
36. Sharpton TJ, Riesenfeld SJ, Kembel SW, Ladau J, O'Dwyer JP, Green JL, Eisen JA, Pollard KS:PhylOTU: A High-Throughput Procedure Quantifies Microbial Community Diversity and Resolves Novel Taxa from Metagenomic Data. *PLoS Comput Biol* 2011, 7:e1001061
37. Zak JC, Willig MR, Moorhead DL, Wildman HG: Functional diversity of microbial communities: A quantitative approach. *Soil Biol Biochem* 1994, 26:1101-1108.
38. Brown JH, Gillooly JF, Allen AP, Savage VM, West GB: TOWARD A METABOLIC THEORY OF ECOLOGY. *Ecology* 2004, 85:1771-1789.
39. Lozupone C, Knight R: UniFrac: a New Phylogenetic Method for Comparing Microbial Communities. *Appl Environ Microbiol* 2005, 71:8228-8235.

40. Valentine DL: Adaptations to energy stress dictate the ecology and evolution of the Archaea. *Nat Rev Micro* 2007, 5:316-323.
41. Altschul SF, Gish W, Miller W, Myers EW, Lipman DJ: Basic local alignment search tool. *J Mol Biol* 1990, 215:403-410.
42. Felsenstein J: PHYLIP - Phylogeny Inference Package (Version 3.2). *Cladistics* 1989, 5:164-166.
43. Stajich JE, Block D, Boulez K, Brenner SE, Chervitz SA, Dagdigian C, Fuellen G, Gilbert JGR, Korf I, Lapp H, Lehväslaiho H, Matsalla C, Mungall CJ, Osborne BI, Pocock MR, Schattner P, Senger M, Stein LD, Stupka E, Wilkinson MD, Birney E: The Bioperl Toolkit: Perl Modules for the Life Sciences. *Genome Res* 2002, 12:1611-1618.
44. Development Core Team: *R: A language and environment for statistical computing*. Vienna, Austria: R Foundation for Statistical Computing; 2005.
45. Dixon P: VEGAN, a package of R functions for community ecology. *J Veg Sci* 2003, 14:927-930.
46. Kanehisa M, Goto S: KEGG: Kyoto Encyclopedia of Genes and Genomes. *Nucl Acids Res* 2000, 28:27-30.

Chapter 2

Determination of Radionuclides and Elemental Composition of Clay Soils by Gamma- and X-ray Spectrometry

Isinkaye M Omoniyi[1], Shitta M B Oludare[1], and Oderinde M Oluwaseyi[2]

[1]Department of Physics, Ekiti State University, P. M. B. 5363, Ado Ekiti, Nigeria

[2]Department of Physics, University of Ibadan, Ibadan, Nigeria

ABSTRACT

Radiochemical and elemental analysis of clay soils collected from different locations within Ekiti State have been performed in this study using gamma and XRF spectrometric measurements. The results of this study show that the mean concentrations of uranium ranged from 2.2 ± 1.0 mg/kg to 3.2 ± 1.1 mg/kg that of thorium ranged from 4.0 ± 0.5 mg/kg to 5.7 ± 1.7 mg/kg, while potasium presented in % by weight ranged from 0.4 ± 0.2 to 1.3 ± 0.3 in all the locations. The overall mean concentrations of these radionuclides are comparable to values from other locations around the world. The XRF analysis revealed 4 major elements and 11 minor or trace elements present in the clay samples. The distribution of the various major and trace elements in all the sampling sites do not follow any systematic trend but vary from point to point. To assess the level of contamination and the possible anthropogenic impact in the clay soils, the enrichment factor (EF) and the geoaccumulation index (Igeo) were estimated for some potential hazardous elements. The results indicate that Cu, Zn, Ni and Mn have $EF < 2$ indicating minimal or no enrichment while Pb is moderately enriched in all the locations.

BACKGROUND

The earth crust contains small amount of uranium, thorium, potassium and other trace and major elements such as Cs, Cd, Pb, Fe, Mg, Mn, etc. The concentrations of all these elements depend on the geology of a local environment as well as other natural and anthropogenic processes. The average concentration of uranium in the earth crust has been reported to be in the range of 2–3 ppm, while thorium exists in the range of 8–12 ppm (IAEA, 2003a). Potassium is widely distributed in nature, with concentrations varying from about 0.1% for limestone, through 1% for sandstones to as much as 3.5% for some granite (Eisenbud, 1987; Eisenbud and Gesell, 1997). The concentrations of major and trace elements in environmental samples had been studied by several authors using either atomic absorption spectrometry (AAS) (Mico et al., 2006; Fagbote and Olanipekun, 2010; Zheng et al., 2010, Ghrefat et al., 2010; Ali and Malik, 2011) or XRF analysis (Kierzek et al., 1999; Boyle,2000; Baranowski et al., 2002; Zhang et al., 2003; Rauf et al., 2004; Bakraji et al., 2010). Most of these studies indicate high concentrations of major and minor elements in the environment. Pollution of natural environment by metals is a worldwide problem because these metals are indestructible and many of them have toxic effects on living organism, especially when they exceed certain threshold (Forstner, 1990; Ghrefat and Yusuf, 2006).

Soil forms a major component of an ecosystem and is the most endangered due to the influence of various human activities such as urban development, industrial and technological advancements, agricultural practices and indiscriminate waste disposal. Soil is considered contaminated when chemicals are present or other alterations have been made to its natural environment (Gowd et al, 2010). Clay is a natural earthy fine grained inorganic material that develops plasticity when mixed with limited amount of water (McGraw-Hill, 1997; Odo et al. 2008). Its origin could be traced to the breaking down of granite rocks by physical and chemical processes called weathering. Natural clay minerals are well known to mankind from the earliest days of civilization and because of their low cost, abundance in most continents of the world, high sorption characteristics and potential for ion exchange, they form a good material for absorbents (Nayak and Singh, 2007). Clay soils generally contain mostly silica (47%) and alumina (40%), elemental analysis have shown that a great number of minor and trace metallic elements such as Sc, Cr, Cu, Ti, Ga, Zr, Mn, Mg, Sr and Pb exist in clay soil. There are several classes of clay which include; smectites (montmorillonite, saponite), mica (illite), kaolinite, serpentine, pylophyllite (talc), vermiculite and sepiolite (Nayak and Singh, 2007). The specific elemental composition of each clay material will usually depend on the amount of the element present in the host rock, the chemical association of the elements with stable and/or unstable mineral during weathering and the intensity of drainage and other polygenetic alterations associated with clay materials (Ibeanu et al., 1997). It has a wide range of applications in the building and ceramic industries. In the rural area, clay is used for the building of earthen or mud houses, while in the urban area, it is used for making burnt bricks used in building modern dwellings. Clay has been in use for making pottery in different continents of the world for several centuries. It is also used in the manufacturing of refractory's products and serves as a natural filter for underground water. In some continents of the world, including Africa, it is a common practice for people to engage in the act of eating clay (geophagia) during pregnancy or for curative purposes. All these may lead to direct or indirect accumulation of metals in man.

XRF is a rapid, non-destructive multi-elemental analysis technique with sensitivity in the range of 10^{-8} (IAEA, 2003b) and it is ideal for environmental research. This analytical method has been widely and routinely applied to the analysis of various archaeological samples, historical relics and works of art (Bakraji et al., 2010; Pillay, 2001; Feretti, 2000). XRF can analyze some 15–30 elements with atomic numbers ranging from Z=11 to Z=41 and some rare earth elements (REEs) (Bakraji et al.,2010). X-ray fluorescence (XRF) analysis is based on the measurement of characteristics X-rays resulting from de-excitation of inner shell vacancy produced in a sample by means of a suitable source of radiation. Energy-dispersive XRF analysis (EDXRF) employs detectors that directly measure the energy of the X-rays by collecting the ionization products in a suitable detecting medium (Tajani and Markowicz, 2004). Gamma ray spectrometry is another analytical technique used widely in environmental investigations. It is used mainly for the determination of the concentrations of radioactive elements that decay through gamma emission. The use of gamma ray spectrometry as a tool for mapping radio elemental concentrations has found widespread acceptance in diverse fields. The method is widely used for environmental monitoring, geological mapping and mineral exploration. It is also a non-radioanalytical technique. Even though, there are many naturally occurring elements that have radioactive isotopes, only potassium and the uranium and thorium decay series, have radioisotopes that produce gamma rays of sufficient energy and intensity measurable by gamma ray spectrometry due to their relative abundance in nature. Geochemical analysis of the total concentrations of major rock-forming elements is important because they provide valuable information about the geochemical properties of the soil in any given environment. The crossed analysis of geochemical and radiometric data provides a useful tool for a better understanding of the origin and characteristics of different rocks (Brai et al. 2006) and soils. It is reported in literature that higher percentage of some metallic oxides in rocks and soil will result in higher specific activities of primordial radionuclides such as ^{238}U, ^{232}Th and ^{40}K in such rocks and soil (Brai et al. 2006). The present work is aimed at evaluating the radionuclides and elemental composition of clay soil using gamma- and X-ray spectrometry. It is also the objective of the study to assess the extent and degree of pollution by metals and identify the origin of the metals using the enrichment factor and geo-accumulation index.

MATERIALS AND METHODS

Sample Collection

The clay samples used for these analyses were collected from five major clay deposits identified within Ekiti State in the south-western part of Nigeria. The study area is situated between longitudes 4° 45″ to 5° 45″ East of the Greenwich Meridian and latitudes 7°15″ to 8° 5″ North of the Equator. The area is completely within the geological basement complex region of Nigeria. In all, 25 samples were analyzed for their radionuclides and elemental concentrations. The samples were collected directly from the exiting mining sites. This gives a good representation of the actual material being utilize either as building material or those used in pottery making. The collected sample were packed into black polythene bags and transported to the laboratory where they were initial air dried at room temperature for about 5 days in order to reduce the moisture content.

Sample Preparation and Measurement for XRF Analysis

To satisfy homogeneity condition of XRF analysis, the clay samples were pulverized manually to very fine powder with an agate mortar and pestle. Pellets of 13 mm diameter were made from 0.3–0.4 g powder without binder at 8 tons of pressure with a hydraulic press. The pellets were kept in different polythene bags which were in turn kept in polypropylene container until analysis. Each sample pellet was irradiated for 1000 seconds at fixed condition of 25 kV and 50 μA. The elemental analysis of the samples was performed using the Energy Dispersive X-ray Fluorescence (EDXRF) spectrometer at the Centre for Energy Research and Development, Obafemi Awolowo University, Ile-Ife, Nigeria. The EDXRF spectrometer consists of a self-contained miniature X-ray tube system ECLIPSE-III, which includes the X-ray tube with a silver (Ag) transmission target, and a beryllium window, a portable Controller incorporating the power supply and control electronics. The Controller generates all the voltages needed to operate the x-ray tube and provides both voltage (kV) and current (μA) display and control. The X-Ray Detector is a Model XR-100CR, high performance thermoelectrically cooled Si-PIN photodiode, with a preamplifier. The detector is powered by the PX2 CR Power supply, which includes a spectroscopy grade Shaping Amplifier. The detector system has a resolution of 220 eV FWHM, for the 5.9 keV peak of ^{55}Fe. The detector is coupled to MCA8000A Multichannel Analyzer for signal processing and data acquisition. The spectrum of Orin sample #5 is shown in Figure 1, while the Logarithm Scale of the same sample is shown in Figure 2. The X-ray tube, ECLIPSE-III with associated Controller/power supply, the Detector system and the Multichannel Analyzer were all supplied by AMPTEK INC., MA USA. The quantitative analysis of the samples was carried out using Fundamental Parameter (FP) method with XRF-FP Software package by CrossRoad Scientific.

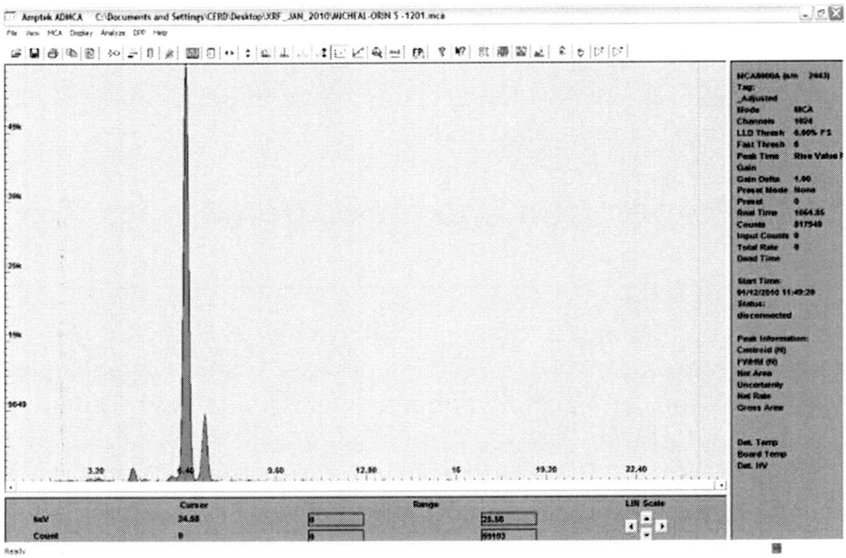

Figure 1: Linear Scale of Orin-5. The two major peaks are K- alpha and beta of iron.

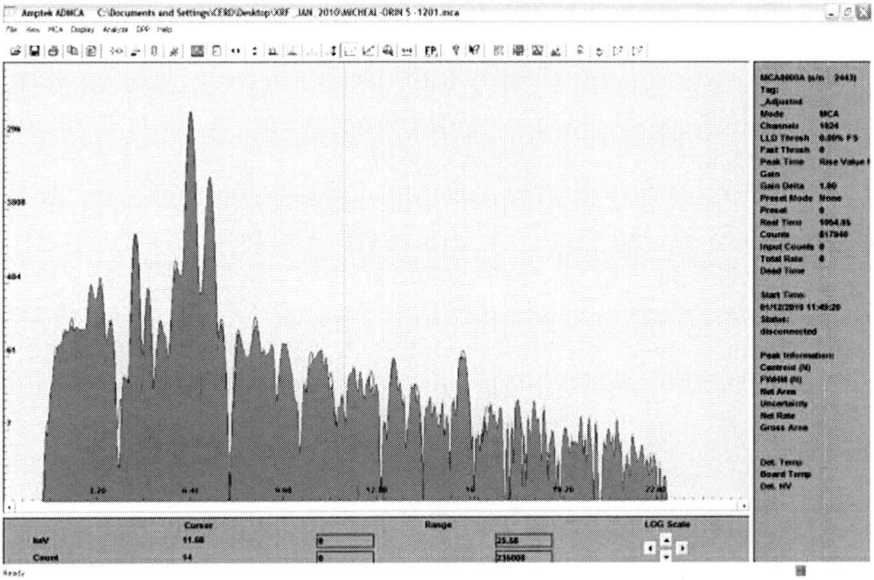

Figure 2: Logarithm Scale of Orin-5.

Sample Preparation and Measurement for Gamma Ray Spectrometric Analysis

Detailed procedure for the preparation and measurement of gamma emitting radionuclides in the clay samples is reported elsewhere (Isinkaye and Shitta, 2010). The air-dried samples were pulverized into powder to achieve uniform matrix similar to the standard sample. All the samples were stored for an upward of 40 d in radon impermeable plastic containers prior to analysis. A 7.6 cm × 7.6 cm NaI(Tl) detector optically coupled to photomultiplier tube was used for the measurement of gamma emitting radionuclides in the clay samples. A multi-channel analyzer matched to an IBM- Compatible personal computer was used for the pulse processing and data analysis. The spectrometer was calibrated against reference material with known activity concentrations of ^{226}Ra, ^{232}Th and ^{40}K (Isinkaye and Shitta, 2010). The detector has a resolution of 8% at the 0.662 MeV line of ^{137}Cs, which is capable of distinguishing the gamma-ray energies of the radionuclides considered in this study. The activity concentrations of

^{226}Ra and ^{232}Th were determined from the gamma lines associated with their respective short-lived daughters; ^{214}Bi (1760 keV) and ^{208}Tl (2615 keV). Each of the samples and the background was counted for 10 h. The background spectral was deducted from the gross count to obtain the net count due to the sample alone.

Evaluation of Massic Elemental Concentrations

The activity concentrations of ^{238}U, ^{232}Th and ^{40}K in Bqkg^{-1} were converted into massic elemental concentrations in part per million (ppm) units for uranium and thorium, and % by weight for potassium, respectively, according to the following equation (Dragovic et al., 2006):

$$F_E = \frac{M_E C}{\lambda_{E.i} N_A f_{E.i}} A_{E.i} \quad (1)$$

where F_E is the fraction of element *E* in the sample, *ME* is the atomic mass (kg mol^{-1}), $_{E,i}$ is the decay constant of the measured isotope of element *E* (s^{-1}), *fE,i* is the fractional atomic abundance in nature, and *AE,i* is the measured specific activity (Bq kg^{-1}) of the radionuclide under consideration (^{238}U, ^{232}Th and ^{40}K), *NA* is the Avogadro's number (6.023×10^{23} atoms mol^{-1}), and *C* is a constant with value of 1,000,000 for U and Th (concentration in ppm) or 100 for K (concentration in % of mass fraction).

RESULTS AND DISCUSSION

The concentrations of 15 major and minor elements together with three radionuclides are presented in Tables 1, 2 and 3. The major elements, Fe, Ti, Ca, and K were estimated as their respective oxides and are given in % by weight, while the minor elements, Cu, Zn, Mn, Zr, Ni, Se, Rb, Sr, Nb, Pb and As are presented in mg/kg unit. The naturally occurring radionuclides; U and Th are presented in ppm units while K is presented in % by weight. The discussion on each group is presented below:

Table 1: Mass concentrations of U, Th and K and external gamma dose rate in clay soil samples

Element	Ado Ekiti	Ara-Ijero Ekiti	Ire Ekiti	Isan Ekiti	Orin Ekiti
	Range	Range	Range	Range	Range
	(Mean ± SD)	(Mean ± SD)	(Mean ± SD)	(Mean ± SD)	(Mean ± SD)
U (ppm)	1.6-4.0	2.0-3.6	1.6-4.1	1.2-3.4	1.5-3.2
	(3.2 ± 1.1)	(2.6 ± 0.7)	(3.1 ± 1.2)	(2.2 ± 1.0)	(2.5 ± 0.8)
232Th (ppm)	3.3-7.2	4.2-6.5	2.7-6.4	3.3-4.5	3.7-8.2
	(5.3 ± 1.6)	(5.6 ± 1.0)	(4.2 ± 1.4)	(4.0 ± 0.5)	(5.7 ± 1.7)
K (%)	1.0-1.6	0.1-0.7	0.2-0.9	0.2-0.8	0.3-1.2
	(1.3 ± 0.3)	(0.4 ± 0.3)	(0.5 ± 0.3)	(0.4 ± 0.2)	(0.7 ± 0.3)
Dose rate (nGyh-1)	30.0-62.3	26.6-39.5	(22.5-43.7)	(20.6-40.7)	(21.8-49.9)
	(48.1 ± 12.2)	(34.6 ± 10.2)	34.8 ± 6.2	27.2 ± 8.4	37.5 ± 11.3

Omoniyi *et al.*

Omoniyi *et al. SpringerPlus* 2013 2:74 doi: 10.1186/2193-1801-2-74

Table 2: Concentrations of major elements in clay soil samples

Element	Ado Ekiti	Ara-Ijero Ekiti	Ire Ekiti	Isan Ekiti	Orin Ekiti
	Range	Range	Range	Range	Range
	(Mean ± SD)	(Mean ± SD)	(Mean ± SD)	(Mean ± SD)	(Mean ± SD)
No of Sample	5	5	5	5	5
Fe2O3 (%)	1.9-17.4	4.9-14.4	3.5-16.4	6.0-22.3	2.8-12.7
	(10.5 ± 7.7)	(11.1 ± 3.9)	(9.7 ± 5.2)	(16.2 ± 6.2)	(7.9 ± 3.9)
TiO2 (%)	1.2-4.7	3.6-5.7	0.8-4.2	6.2-10.4	1.3-4.2
	(3.2 ± 1.7)	(4.9 ± 0.9)	(2.8 ± 1.2)	(7.6 ± 1.6)	(2.9 ± 1.4)
CaO (%)	0.9-6.5	1.0-3.5	0.2-2.5	2.3-4.7	0.2-5.0
	(3.9 ± 2.4)	(2.3 ± 1.0)	(1.2 ± 0.9)	(3.3 ± 0.9)	(1.6 ± 1.3)
K2O (%)	1.2-9.9	3.8-9.2	0.6-5.8	5.3-9.0	0.7-9.5
	(3.2 ± 1.7)	(5.8 ± 2.0)	(2.8 ± 2.0)	(6.3 ± 1.5)	(4.4 ± 3.3)

Omoniyi *et al.*

Omoniyi *et al. SpringerPlus* 2013 **2**:74 doi: 10.1186/2193-1801-2-74

Table 3: Concentrations of minor or trace elements in clay soil samples

Element	Ado Ekiti	Ara-Ijero Ekiti	Ire Ekiti	Isan Ekiti	Orin Ekiti
	Range	Range	Range	Range	Range
	(Mean ± SD)	(Mean ± SD)	(Mean ± SD)	(Mean ± SD)	(Mean ± SD)
Cu (ppm)	32-165	28-190	22-176	18-195	17-174
	(100.6 ± 61.6)	(107.6 ± 63.1)	(91.6 ± 60.8)	143.2 ± 71.9	(82.6 ± 60.9)
Zn (ppm)	90-246	88-290	74-364	95-311	67-246
	(161.0 ± 71.9)	(184.2 ± 76.0)	(202.4 ± 124.3)	257.2 ± 92.5	(166.4 ± 84.7)
Mn (ppm)	219-2823	605-2539	452-3606	780-5296	710-2000
	(1593.0 ± 1190.4)	(1869.4 ± 785.2)	(1682.8 ± 1306.7)	3780.6 ± 1870.7	(1451.4 ± 596.8)
Zr (ppm)	151-953	166-813	46-270	222-486	105-1439
	(443.2 ± 314.5)	(389.0 ± 252.4)	150.2 ± 95.0	332.4 ± 95.7	(439.0 ± 56.4)
Ni (ppm)	88-194	32-269	33-336	27-216	22-273
	(136.8 ± 43.5)	(151.8 ± 95.0)	180.4 ± 128.0	133.2 ± 69.5	(147.2 ± 93.5)
Se (ppm)	23-68	49-97	44-61	43-98	42-101
	(49.0 ± 18.9)	(65.5 ± 21.0)	54.8 ± 8.1	69.5 ± 23.3	(63.3 ± 25.9)

Rb (ppm)	33-104	171-234	30-234	169-282	21-188
	(63.0±31.6)	(202.8±30.4)	107.0±90.4	225.5±50.2	(117.0±74.2)
Sr (ppm)	13-89	91-118	10-90	121-346	23-256
	(57.8±31.9)	(106.8±12.0)	36.6±31.7	225.0±80.4	(91.0±38.8)
Nb (ppm)	35-120	34-94	28-99	54-114	24-79
	(78.3±41.5)	(76.3±28.2)	63.8±29.1	90.5±28.0	(51.8±26.8)
Pb (ppm)	73-499	75-310	72-311	86-317	70-232
	(232.2±164.5)	(199.8±100.9)	174.4±118.0	247.2	(150.6±76.6)
As (ppm)	137-183	ND	ND	ND-49	ND
	(160.0±32.5)			23.0±22.7	

Omoniyi *et al.*

Omoniyi *et al. SpringerPlus* 2013 2:74 doi: 10.1186/2193-1801-2-74

Naturally Occurring Radionuclides

The massic elemental concentrations of three naturally occurring radionuclides U, Th and K measured in the clay soil samples investigated in this study are presented in Table 1. Ado Ekiti samples have the highest mean concentration of uranium and potassium, and the highest mean concentration of thorium is found in Orin Ekiti. All the lowest mean concentrations are obtained in Isan Ekiti. The concentrations of the three radionuclides, U, Th and K, ranged from 1.2–4.1 ppm, 2.7–8.2 ppm and 0.1–1.6%, respectively. These radionuclides showed a range of concentrations, as a consequence of varying geological composition of the studied area.

Massic conconcentrations of U, Th and K in comparison with other locations around the world is presented in Table 4. The mean U (ppm) concentration obtained in this study is greater than values obtained in Albenia, Australia, Cyprus and Italy, lower than values obtained in Canada, Egypt and Jordan but comparable to values obtained in Serbia and Montenegro, and USA (Table 4). The mean concentration of Th (ppm) is higher than those obtained in Cyprus and Egypt but lower than those obtained in Albenia, Australia, Bulgaria, Canada, Italy, Jordan, Serbia and Montenegro, and USA. The mean concentration of K (%) is however lower than the values obtained in all these country except Cyprus. All the radionuclides concentrations except uranium are lower than the world average values (Table 4).

Table 4: The comparison of the mass concentrations of U, Th and K in clay soil samples of the present study with other locations around the world

Location		U (ppm)	Th (ppm)	K (%)	Reference
Albania*		0.48-7.68 (1.84)	0.98-39.5 (5.93)	0.05-3.75 (1.17)	UNSCEAR (2000)
Australia	Rock	2.1-3.6 (2.5)	18-55 (15)	2.4-3.8 (2.4)	Dickson and Scott (1997)
	Soil	1.6-3.8 (2.2)	6-19 (1.3)	0.7-1.9 (1.3)	
Bulgaria*		0.64-15.2 (3.2)	1.73-39.5 (7.41)	0.05-3.75 (1.17)	UNSCEAR (2000)

Canada (Rock)	0.8-16.4 (4.1)	1.1-41.0 (11.9)	1.0-6.2 (3.1)	Killeen (1979)
Cyprus*	0.08-7.2 (1.2)	0.25-13.1 (2.72)	0.04-2.91 (0.58)	Tzortzi et al. (2003)
Italy*	0.16-5.6	0.25-16.8 (5.43)	0.03-5.14 (1.41)	Chiozzi et al. (2002)
Egypt*	0.48-9.6 (2.96)	0.49-23.7 (4.45)	0.09-2.12 (1.04)	UNSCEAR (2000)
Jordan*	1.76-8.32 (6.72)	5.18-25.4 (20.2)	0.45-1.96 (1.82)	Al-Jundi et al. (2003)
Serbia and Montenegro	1.2-6.24 (2.76)	4.45-21.0 (10.4)	0.88-2.99 (1.98)	Dragovic et al. (2006)
USA*	0.32-11.2	0.98-32.1 (8.20)	0.32-2.28 (1.21)	Myrick et al. (1983)
World Average	2.64	11.1	1.37	Dragovic et al. (2006)
Present study	1.2-4.1 (2.7)	2.7-8.2 (4.9)	0.1-1.63 (0.7)	

*Sourced from Dragovic et al. (2006).

Omoniyi *et al.*

Omoniyi *et al. SpringerPlus* 2013 2:74 doi: 10.1186/2193-1801-2-74

Estimation of Gamma Dose Rate

The radiological implications of the activity concentrations of the naturally occurring radionuclides present in the clay samples were estimated using the external gamma dose rates at 1 m above an infinite homogeneous soil medium per unit radioelement concentration assuming radioactive equilibrium in the uranium and thorium decay series. In the calculation, the contributions of artificial radionuclides such as ^{137}Cs and ^{90}Sr were neglected. The calculations were performed according to the following equation (IAEA, 2003a; Lovborg, 1984):

$$D = [13.078A_K + 5.675A_U + 2.494A_{\mathrm{Th}}]\mathrm{nGyh}^{-1} \quad (2)$$

Where, A_K is the mass concentration of K in %, A_U is the mass concentration of uranium in ppm and A_{Th} is the mass concentration of thorium in ppm. The estimation is based on the assumption that

1%K corresponds to 13.078 $nGyh^{-1}$, 1 ppmU gives 5.675 $nGyh^{-1}$ and 1 ppmTh is equivalent to 2.494 $nGyh^{-1}$dose rate, respectively. The range and mean dose rates obtained for all the clay samples in the study locations are presented in Table 1. All the mean gamma dose rates obtained for the five study locations are lower than the world average value of 59 $nGyh^{-1}$ (UNSCEAR, 2000). The effective dose was also estimated using the formula suggested by Dragovic et al., (2006):

$$H_E = [D \times 24 \times 365 \times 0.7 \times 0.2] \quad (3)$$

Where D is the gamma dose rate obtained from mass concentrations of U, Th and K, 0.7 $SvGy^{-1}$ is the conversion coefficient from absorbed dose in air to effective dose and 0.2 represents the outdoor occupancy factor, which shows that the people in the study area spend ~20% of their time outdoor. The mean annual effective dose obtained in the study area varied from 0.03–0.06 mSv, which fall below the worldwide mean annual effective dose value of 0.07 mSv. The results obtained indicate that the study area can be categorized as area with normal background radiation.

Major Elements

The lowest mean concentration value (7.9 ± 3.9%) of Fe_2O_3 obtained in this study is found in Orin-Ekiti while the highest mean concentration value of 16.2 ± 6.2% is obtained in Isan-Ekiti. The values obtained in all sampling points ranged from 1.9–22.3%. The mean concentrations of Fe_2O_3 in all the five locations are higher than the average crustal value reported in Turekian and Wedepohl (1961). The values are also higher than mean value obtained in a Mediterranean agricultural soil in Spain. The high concentration of Fe in the clay soils is generally not of any major concern because Fe is not a contaminant element. Fe is important in plant nutrition and an essential crop micronutrient. The mean concentrations of TiO_2 follow the same trend as Fe_2O_3 with lowest mean value (2.9 ± 1.4%) obtained in Orin-Ekiti while the highest mean value of 7.9 ± 1.6% is obtained in Isan-Ekiti. TiO_2 is the most common compound of Titanium and is widely distributed in the Earth's crust. It is found in almost all living things, rocks, water bodies and soil

(Wikipedia, 2011). Its proportion in soil is approximately 0.5–1.5% (Barksdale, 1968). The concentration of CaO varied from 0.2–6.5% with highest mean concentration found in Ado-Ekiti and the lowest concentration obtained in Ire-Ekiti. Similarly, the concentrations of K_2O vary from 0.6–9.9% with highest and lowest mean concentrations of 6.3 ± 1.5% and 2.8 ± 2.0% respectively. The Ca and K concentrations obtained in this study are comparable to the values obtained by XRF analysis of some clay samples in Pakistan (Baranowski et al., 2002). Their results indicate a range of Ca to be 0.10–8.9% while K ranged from 0.05–2.25%. Also the levels of Ca and K obtained in this study is higher than mean concentrations of 0.35% and 0.24% obtained respectively for Ca and K in coal samples by Kierzek et al. (1999).

Minor Elements

The results showed that Mn has the highest overall mean concentration, followed by Zr, Pb, Zn, Sr, Nb, Se, As (Table 3). All the sampling locations showed higher Cu, Zn, Mn, Ni, Pb and As contents than the values obtained for average shale as reported by Turekian and Wedepohl (1961). As seen, the distributions of these minor elements vary much from different sampling locations i.e the distribution does not follow any systematic trend. For Cu, the mean concentrations vary from 82.6 ± 60.9-143.2 ± 71.9 mg/kg, Zn, 166.4 ± 84.7-257.2 ± 92.5 mg/kg, Mn, 1451.4 ± 596.8-3780.6 ± 1870.7 mg/kg, Zr, 150.2 ± 95.0-443.2 ± 314.5 mg/kg, Ni, 133.2 ± 69.5-180.4 ± 128.0 mg/kg, Se, 49.0 ± 18.9-69.5 ± 23.3 mg/kg, Rb, 63.0 ± 31.6-225.5 ± 50.2 mg/kg, Sr, 36.6 ± 31.7-225.0 ± 80.4 mg/kg, Nb, 51.8 ± 26.8-90.5 ± 28.0 mg/kg, Pb, 150.6 ± 76.6-247.2 ± mg/kg, and As, 23.0 ± 22.7-160.0 ± 32.5 mg/kg. The high standard deviation values indicate the spread in the distribution of the minor elements in all the sampling sites. Most of the minor elements have their highest mean concentrations at Isan-Ekiti sampling sites. Isan-Ekiti clay is kaolinitic in nature and it is used extensively in making earthen vessels used by local populace for cooking. This could pose metal poisoning and some other detrimental health hazard. Some of the potentially hazardous element such as Cu, Zn, Ni, Pb and As have their mean concentrations higher than the maximum allowable concentration values in clay soil as applied in the Federal Republic of Germany. For Cu, Zn, Ni and Pb, the maximum allowable concentration in clay soil are respectively, 60 mg/kg, 200 mg/kg, 70 mg/kg and 100 mg/kg.

Enrichment Factor and Geoaccumulation Index

In order to assess the level of contamination and the possible anthropogenic impact in the clay samples, the enrichment factors (EF) and geoaccumulation index (I_{geo}) were estimated for some selected potentially hazardous elements evaluated in this study. The enrichment factor is defined as:

$$EF = \left(\frac{C_x}{C_{Fe}}\right)_{sample} \Big/ \left(\frac{C_x}{C_{Fe}}\right)_{refence} \quad (4)$$

Where C_x is the concentration of the potentially enrichment element and C_{Fe} is the concentration of the proxy or normalizing element usually Fe. The world average elemental concentrations reported by Turekian and Wedepohl (1961) in the earth's crust were used as reference in this study because regional geochemical background values for these elements are not available. Five major contamination categories are recognize on the basis of the enrichment factor, where, $EF < 2$ is deficient to minimal enrichment, $2 \leq EF < 5$ is moderate enrichment, $5 \leq EF < 20$ is significant enrichment, $20 \leq EF < 40$ means high enrichment, and $EF > 40$ indicates extremely high enrichment. EF can easily be used to differentiate between elemental concentrations from anthropogenic source and those from natural origin. According to Zhang and Liu (2002), EF values between 0.5 and 1.5 indicate the metal is entirely from crustal materials or natural origin, while $EF > 1.5$ suggests that the sources are more likely to be anthropogenic (Ghrefat et al., 2010). The results of the present study show EF values of Cu, Zn, Ni and Mn which are all < 2(Table 5), indicating no or minimal enrichment. Pb is moderately enriched in all the locations while as is moderately enriched only in Ado-Ekiti clay samples. All the potentially hazardous elements considered in the study originate from the source rock except Pb which has $EF > 1.5$, indicating anthropogenic source.

Table 5: Average concentrations of selected elements, average shale (ppm) (Turekian and Wedepohl, 1961), enrichment factor and the geoaccumulation index values in the clay samples

Elements	Average Shale Value	Ado Ekiti	EF	Igeo	Ara-Ijero Ekiti	EF	Igeo	Ire Ekiti	EF	Igeo	Isan Ekiti	EF	Igeo	Orin Ekiti	EF	Igeo
Cu	45	100.6	1.00	0.58	107.6	1.01	0.67	91.6	0.99	0.44	143.2	0.92	1.09	82.6	1.09	0.29
Zn	95	161	0.76	0.18	184.2	0.82	0.37	202.4	1.03	0.51	257.2	0.79	0.85	166.4	1.04	0.22
Ni	68	136.8	0.90	0.42	151.8	0.95	0.57	180.4	1.29	0.82	133.2	0.57	0.39	147.2	1.29	0.53
Mn	850	1593	0.84	0.32	1869.4	0.93	0.55	1682.8	0.96	0.40	3780.6	1.29	1.57	1451.4	1.02	0.19
Pb	20	232.2	5.20	2.95	199.8	4.23	2.74	174.4	4.23	2.54	247.2	3.59	3.04	150.6	4.48	2.33
As	13	160	5.51	3.04	-	-	-	-	-	-	23	0.51	-	-	-	-

Omoniyi *et al.*

Omoniyi *et al. SpringerPlus* 2013 2:74 doi: 10.1186/2193-1801-2-74

The geo-accumulation index (Igeo) originally introduced and applied by Muller (1969) was used to evaluate the degree of elemental pollution in the clay soils from the study area. Mathematically, Igeo is given as (Zheng et al., 2010; Matini et al., 2011):

$$I_{\text{geo}} = \text{Log}_2\left(\frac{C_n}{\text{kB}_n}\right) \quad (5)$$

Where C_n is the concentration of the potentially hazardous trace element (e.g Cu, Ni, Pb, etc) in the clay sample, B_n is the geochemical background value in average shale (Turekian and Wedepohl,1961) of element n and $k = 1.5$ is the background matrix correction factor introduced to account for possible differences in the background values due to lithogenic effects. The results of the geo-accumulation index obtained in this study indicate that Cu is moderately contaminated in Isan Ekiti clay with $Igeo = 1.09$(Tables 5 and 6). Pb is moderately/strongly contaminated in all the sampled locations. The anthroponenic sources of Pb include; exhaust fumes from motor-vehicle, smelting activities, indiscriminate dumping of used lead acid batteries, etc.

Table 6: Geoaccumulation index (Igeo) for contamination levels in clay soil samples

Igeoclass	Igeovalue	Contamination level
0	Igeo ≤ 0	Uncontaminated
1	0 < Igeo ≤ 1	Uncontaminated/moderately contaminated
2	1 < Igeo ≤ 2	Moderately contaminated
3	2 < Igeo ≤ 3	Moderately/strongly contaminated
4	3 < Igeo ≤ 4	Strongly contaminated
5	4 < Igeo ≤ 5	Strongly/extremely contaminated
6	Igeo > 5	Extremely contaminated

Omoniyi *et al.*

Omoniyi *et al. SpringerPlus* 2013 2:74 doi: 10.1186/2193-1801-2-74

Statistical Analysis

Table 7 gives the descriptive statistics for the massic elemental concentrations of U, Th and K for all the measured clay samples. These includes; arithmetic means, median, standard deviation, maximum, minimum, skewness and Kurtosis, while the frequency distributions are presented in Figure 3.

Table 7: Descriptive statistics of massic elemental concentrations of U, Th and K for all the analyzed clay samples

Parameter	Massic elemental concentration		
	U (ppm)	Th (ppm)	K (%)
Mean	2.7212	4.9229	0.6716
Standard deviation	0.96973	1.40750	0.42847
Maximum	4.11	8.20	1.63
Minimum	1.17	2.73	0.09
Median	2.6215	4.5394	0.6670
Skewness	−0.005	0.564	0.518
Kurtosis	−1.366	−0.384	−0.596

Omoniyi *et al.*

Omoniyi *et al. SpringerPlus* 2013 2:74 doi: 10.1186/2193-1801-2-74

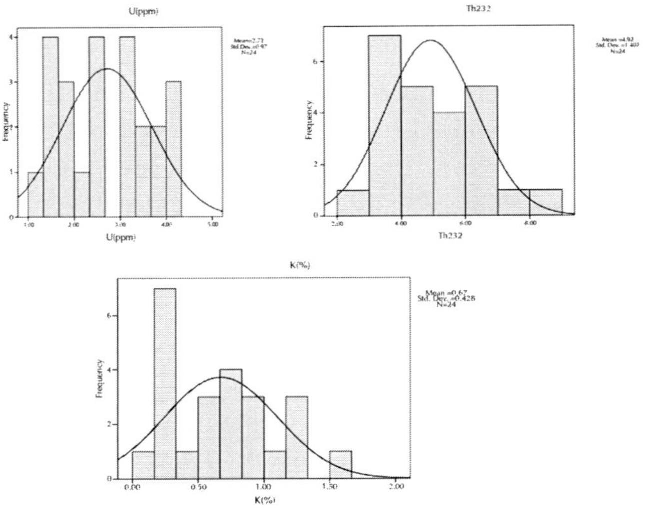

Figure 3: Frequency distribution of the mass concentrations of U, Th and K in clay soil.

The results of the Pearson correlation coefficients between the naturally occurring radionuclides and the major and trace elements are presented in Table 8. The results indicate a strong positive correlation between radioactive K and As. This radioactive K has a poor negative correlation with almost all the other elements except Ca, Se, Nb and Pb. U correlates significantly only with Th and Sr at 95% confidence level. Th does not interact significantly with any of the major and minor elements. Fe correlates significantly with Ti, Cu, Zn, Mn, Ni, Se, Rb, Nb and Pb, whereas Ti has strong interactions with Cu and Mn. There exists a strong correlation between Ca and As; K and Sr; Cu and Zn, Mn, Ni, Se, Rb, Nb, Pb; Zn and Mn, Ni, Se, Rb, Nb, Pb; Mn and Se, Rb, Nb, Pb. Ni correlates strongly with Se, Rb, Nb, Pb while Se has a strong correlates with Rb, Nb,Pb. Strong interactions also exist between Pb and As. All the three radionuclides considered in this study are poorly correlated with the measured major and trace elements indicating different geochemical behavior. Positive correlations however exist between most of the elemental pairs in the clay samples suggesting the same source or co-contamination. Negative or inverse correlations between variables indicate that the variable pairs are derived from different origin and that such do not associate in their geochemical dynamics.

Table 8: Pearson correlation coefficients between natural radionuclides and elemental concentrations in clay soil samples

Variable	K(%)	U(ppm)	Th(ppm)	Fe	Ti	Ca	K	Cu	Zn	Mn	Zr	Ni	Se	Rb	Sr	Nb	Pb	As
K(%)	1																	
U(ppm)	.499*	1																
Th(ppm)	.299	.426*	1															
Fe	-.022	.079	.161	1														
Ti	-.292	-.296	.001	.574**	1													
Ca	.170	-.194	-.019	.406*	.215	1												
K	-.472*	-.366	.103	.157	.244	.486*	1											
Cu	-.020	.030	.173	.910**	.592**	.229	-.007	1										
Zn	-.165	-.075	.105	.864**	.499*	.113	.001	.906**	1									
Mn	-.176	-.071	.014	.913**	.577**	.272	.184	.866**	.858**	1								
Zr	-.100	-.085	.015	-.198	-.096	.406*	.389	-.263	-.292	-.139	1							
Ni	-.056	.074	.291	.550**	.239	-.172	-.264	.762**	.775**	.501*	-.369	1						
Se	.121	.186	.342	.590**	.153	-.140	-.116	.639**	.651**	.616**	-.496*	.534**	1					
Rb	-.252	-.113	.084	.765**	.446*	-.016	.104	.758**	.823**	.796**	-.305	.546**	.753**	1				
Sr	-.446*	-.452*	-.194	.216	.454*	.503*	.695**	.062	.077	.317	.417*	-.258	-.149	.165	1			
Nb	.065	.059	.139	.806**	.328	.204	-.192	.835**	.827**	.779**	-.297	.677**	.671**	.717**	-.104	1		
Pb	.161	.248	.383	.839**	.397*	.333	-.049	.843**	.783**	.741**	-.149	.678**	.622**	.600**	.031	.816**	1	
As	.570**	.224	.147	.414*	.067	.591**	-.310	.343	.215	.223	.075	.145	.062	-.016	-.099	.430*	.564**	1

*Significant at the 0.05 level.

**Significant at the 0.01 level.

Omoniyi *et al.*

Omoniyi *et al.* *SpringerPlus* 2013 2:74 doi: 10.1186/2193-1801-2-74

CONCLUSIONS

Radiometric and elemental investigations performed on clay soils reveal the presence of three naturally occurring radionuclides and fifteen major and trace elements. The results show that the radionuclides and elemental concentrations varied widely among the sampling locations. Most of the elements have higher concentrations than the baseline values. The concentrations of U (ppm), Th (ppm) and K (%) are comparable with results from other locations around the world and lower than the world average except U. Soil pollution in the present study was assessed using enrichment factor and geoaccumulation index values. The results indicate that the clay soil samples examined in this study are unpolluted with Cu, Zn, Ni, and Mn. Pb is moderately contaminated in all the sampling locations. The study shows that all the potentially hazardous elements originate from the source rock except Pb, which has EF > 1.5, indicating anthropogenic source.

Descriptive statistics and correlation analysis was carried out on the results in order to have a better understanding of the complex dynamics of the measured parameters. The Pearson correlation analysis shows poor interactions between radionuclides and elemental concentrations. Strong positive correlations were observed among most elemental pairs suggesting the same origin and similar geochemical behavior.

AUTHORS' CONTRIBUTION

IMO carried out the sample collection and drafted the manuscript. SMBO edited the drafted manuscript.OMO carried out the sample analysis. All authors read and approved the final manuscript.

REFERENCES

1. Ali SM, Malik RN (2011) Spatial distribution of metals in top soils of Islamabad City, Pakistan. Environ Monit Assess 172:1-16
2. Al-Jundi J, Al-Bataina BA, Abu-Rukah Y, Shchadch HM (2003) Natural radioactivity concentrations in soil samples along the Amman Aqaba Highway, Jordan. Radiat Meas 36:555-560

3. Bakraji EH, Itlas M, Abdulrahman A, Issa H, Abboud R (2010) X-ray fluorescence analysis for the study of fragments pottery excavated at Tell Jendares site, Syria, employing multivariate statistical analysis. J Radioanal Nucl Chem 285:455-460
4. Baranowski R, Rybak A, Baranowska I (2002) Speciation Analysis of Elements in Soil Samples by XRF. Polish J Environ Stud 11(5):473-482
5. Barksdale J (1968) "Titanium". In: Hampel CA (ed) The Encyclopedia of the Chemical Elements, Reinhold Book Corporation, New York. pp 732-738 LCCN 68-29938
6. Boyle JF (2000) Rapid elemental analysis of sediment samples by isotope source XRF. J Paleol 23:213-221
7. Brai M, Belli S, Hauser S, Puccio P, Rizzo S, Basile S, Marrale M (2006) Correlation of radioactivity measurements, air kerma rates and geological features of Sicily. Radiat Meas 41:461-470
8. Chiozzi P, Pascale V, Verdoya M (2002) Naturally occurring radioactivity at the Alps-Apennines transition. Radiat Meas 35:147-154
9. Dickson BL, Scott KM (1997) Interpretation of aerial gamma ray surveys-adding the geochemical factors. AGSO J Australia Geol Geophys 17(2):187-200
10. Dragovic S, Lj J, Onjia A, Bacic G (2006) Distribution of primordial radionuclides in surface soils from Serbia and Montenegro. Radiat Meas 41:611-616
11. Eisenbud M (1987) Environmental radioactivity. Academic Press, Orlando, USA.
12. Eisenbud M, Gesell T (1997) Environmental radioactivity from natural, industrial and military sources. Academic Press, San Diego, California, USA.
13. Fagbote EO, Olanipekun EO (2010) Evaluation of the status of heavy metal pollution of soil and plant (Chromolaena odoranta) of Agbabu bitumen deposit area, Nigeria. American-Eurosian J Sci Res 5(4):241-248
14. Feretti M (2000)Creagh DC, Bradley DA (eds) Radiation in art and archaeometry, Elsevier, Amsterdam. p 285
15. Forstner U (1990) Contaminated sediments. Lecture Notes in Earth Science, vol 21. Springer-Verlag, Berlin.

16. Ghrefat HA, Yusuf N (2006) Assessment Mn, Fe, Cu, Zn, and Cd pollution in bottom sediments of Wadi Al-Arab Dam. Jordan. Chemosphere 65:2114-2121

17. Ghrefat HA, Abu-Rukah Y, Rosen MA (2010) Application of geoaccumulation index and enrichment factor for assessing metal contamination in the sediments of Kafrain Dam, Jordan. Environ Monit Assess.

18. Gowd SS, Reddy M, Govil PK (2010) Assessment of heavy metal contamination in soils at Jajmau (Kanpur) and Unnao industrial areas of the Ganga Plain, Uttar Pradesh, India. J Hazard Mat 174:113-121

19. IAEA (2003a) International Atomic Energy Agency. Guidelines for radioelement mapping using gamma ray spectrometry data. IAEA-TECDOC-1363. IAEA, Vienna, Austria.

20. IAEA (2003b) International Atomic Energy Agency. Collection and preparation of bottom sediment samples for analysis of radionuclides and trace elements. IAEA-TECDOC-1360. IAEA, Vienna, Austria.

21. Ibeanu IGE, Dim LA, Mallam SP, Akpa TC, Munithya J (1997) Non-Destructive XRF Analysis of Nigerian and Kenyan Clays. J Radioanal Nucl Chem 221(1–2):207-209

22. Isinkaye MO, Shitta MBO (2010) Natural radionuclide content and radiological assessment of clay soils collected from different sites in Ekiti State, southwestern Nigeria. Radiat Prot Dosim 139(4):590-596

23. Kierzek J, Malozewska-Bucko B, Bukowski P, Parus JL, Ciurapinski A, Zaras S, Kunach B, Wiland K (1999) Assessment of coal and ash environmental impact with the use of gamma- and X-ray spectrometry. J Radioanal Nucl Chem 240(1):39-45

24. Killeen PG (1979) Gamma ray spectrometric methods in uranium exploration – Application and interpretation. In: Geophysics and Geochemistry in the Search for Metallic Ores, edited by PJ Hood. Geophysical Survey of Canada Economic Geology Report 31:163-230

25. Lovborg L (1984) The calibration of portable and airborne gamma ray spectrometers- theory, problems and facilities. Report Riso-M-2456, Roskilde.

26. Matini L, Ongoka PR, Tathy JP (2011) Heavy metals in soil on spoil heap of an abandoned lead ore treatment plant, SE Congo-Brazzaville. African J Environ Sci Tech 5(2):89-97
27. McGraw-Hill (1997) Encyclopedia of Science and Technology; 15th edn vol 3. McGraw-Hill Book Company, New York.
28. Mico C, Peris M, Sanchez J, Recatala L (2006) Heavy metal content of agricultural soils in a Mediterranean semiarid area: the Segural River valley (Alicante, Spain). Spanish J Agric Res 4(4):363-372
29. Muller G (1969) Index of geoaccumulation in sediments of the Rhine River. Geol J 2:109-118
30. Myrick TE, Berven BA, Haywood FF (1983) Determination of concentration of selected radionuclides in surface soil in the U.S. Health Phys 45:631-642
31. Nayak PS, Singh BK (2007) Instrumental characterization of clay by XRF, XRD and FTIR. Bulletin Mat Sci 30(3):235-238
32. Odo JU, Mba AC, Udenya TC (2008) Effect of agricultural waste ash additives on refractory properties of a blend of two Nigerian clays. J Metallurgy Mat Eng 3(1):30-34
33. Pillay AE (2001) Analysis of archaeological artefacts: PIXE, XRF or ICP-MS? J Radioanal Nucl Chem 247(3):593-595
34. Rauf MA, Ikram M, Iqbal MJ, Manzoor S (2004) Comparizon of catalytic activity of clays on locally availablepetroleum fractions. J Chem Soc Pakistan 26(1):10-13
35. Tajani A, Markowicz A (2004) EDXRF analysis of thin samples. In: IAEA-TECDOC-1401. Quantifying uncertainty in nuclear analytical measurements. International Atomic Energy Agency (IAEA), Vienna, Austria.
36. Turekian YY, Wedepohl KH (1961) Distribution of the elements in some major units of the earth's crust. Geol Soc America 72:175-192
37. Tzortzi M, Tsertos H, Christisfides S, Christodoulides G (2003) Gamma-ray measurements of naturally occurring radioactive samples from Cyrus characteristic geological rocks. Radiat Meas 37:221-229
38. UNSCEAR (2000) United Nations Scientific Committee on Effects of Atomic Radiation (UNSCEAR), Sources and Effects of Ionizing Radiation. United Nations, New York. Annex A&B

39. Wikipedia (2011) The Free Encyclopedia. http://en.wikipedia.org/wiki/Titanium. Retrieved on 25th Oct. 2011
40. Zhang J, Liu CL (2002) Riverine composition and estuarine geochemistry of particulate metals in China-Weathering features, anthropogenic impact and chemical fluxes. Estuarine Coastal Shelf S 54:1051-1070
41. Zhang Y, Erhkang L, Deyi L, Yinsong W, Yuehchang Y, Changwan W, Waiguo S, Min Z, Guilin Z, Yan L (2003) PIXE and radioactivity measurements for elemental determination in river water and sediment samples. J Radioanal Nucl Chem 258(2):415-419
42. Zheng LG, Liu GJ, Kang Y, Yang RK (2010) Some potential hazardous trace elements contamination and their ecological risk in sediments of western Chaohu Lake, China. Environ Monit Assess 166:379-386

Chapter 3

Review of Underground Coal Gasification Technologies and Carbon Capture

Stuart J Self, Bale V Reddy, and Marc A Rosen

Faculty of Engineering and Applied Science, University of Ontario Institute of Technology, 2000 Simcoe Street North, Oshawa, ON, L1H 7K4, Canada

ABSTRACT

It is thought that the world coal reserve is close to 150 years, which only includes recoverable reserves using conventional techniques.

Mining is the typical method of extracting coal, but it has been estimated that only 15% to 20% of the total coal resources can be recovered in this manner. If unrecoverable coal is considered in the reserves, the lifetime of this resource would be greatly extended, by perhaps a couple hundred years. Mining involves a large amount of time, resources, and personnel and contains many challenges such as drastic changes in landscapes, high machinery costs, elevated risk to personnel, and post-extraction transport. A new type of coal extraction method, known as underground coal gasification (UCG), that addresses most of the problems of coal mining is being investigated and implemented globally. UCG is a gasification process applied to *in situ* coal seams. UCG is very similar to aboveground gasification where syngas is produced through the same chemical reactions that occur in surface gasifies. UCG has a large potential for providing a clean energy source through carbon capture and storage techniques and offers a unique option for CO_2 storage. This paper reviews key concepts and technologies of underground coal gasification, providing insights into this developing coal conversion method.

REVIEW

Introduction

The global energy supply is comprised of many different sources, including fossil fuels, uranium, and various alternative and renewable sources. Currently, over 85% of the global energy supply is derived from fossil fuels, and a high fossil fuel dependency appears likely to remain in the immediate future [1]. The amount of energy required globally is projected to increase due to growing population and industrialization [2, 3]. Some feel that the total primary energy demand will double or even triple by the year 2050 relative to levels today, and as the energy demand continues to increase, future fossil fuel shortages are predicted [4,5].

Hammond [6] argues that fossil fuel depletion is a significant factor when considering sustainable energy systems for the future. Fossil fuel resources are finite and being consumed rapidly, beginning with the most economically attractive resources [7]. In the future, fossil fuel resource extraction and production rates are expected to peak and begin to decline [7]. Oil production is predicted to peak in 5 to 15 years and gas production within 40 years, with significant exhaustion of oil and gas reserves by the years 2050 and 2070, respectively [4,8]. As fossil fuel demand approaches supply levels, the cost of energy is anticipated to increase drastically, prompting research and technological developments for improved ways to convert more fossil fuel resources into useable reserves [9].

Currently, coal generates 41.5% of the world's electricity and provides 26.5% of global primary energy needs [1]. Coal has the largest reserves in the world of the fossil fuels and is abundant in many countries. It is thought that the world's recoverable coal reserve is close to 150 years at current production rates, but this only represents 15% to 20% of the entire resource [8]. Remaining global coal resources have recently been estimated to be 18 trillion tonnes [10]. This contrasts significantly with the typical figure of tens of billions of tonnes for recoverable reserves [4]. If unrecoverable coal is considered in the recoverable reserves, the lifetime of the resource could be extended by a couple hundred years. For this to be realized, new, economic extraction techniques need to be implemented.

Coal is conventionally extracted by mining, both underground and open pit. Mining operations require much time, personnel, and natural resources; typically, coal reserves lie too deep underground, or are otherwise too costly, to exploit using conventional mining methods. Conventional mining also has other issues including land subsidence, high machinery costs, hazardous work environments, coal transport requirements, localized flooding, and methane buildup in cellars of nearby homes [11].

Underground coal gasification (UCG) is a newer type of coal extraction that is being investigated and implemented around the world and that avoids most of the problems of mining coal. UCG involves the conversion of unmanned coal, where coal seams are gasified, without mining, and synthetic gas (syngas) is produced for use in power generation or as chemical feedstock [12]. UCG limits the amount of underground work required by personnel, lowering risks of harm relative to conventional mining. Power generation and chemical processing plants can be built directly above a coal resource and use syngas produced through UCG, avoiding coal transport. UCG has the ability to significantly widen the resource base, where the energy contained within inaccessible coal reserves, considered uneconomical for recovery, could be recovered using UCG [11]. It has been estimated by the Underground Coal Gasification Partnership that around 4 trillion tonnes of otherwise unusable coal could be suitable for UCG [9].

UCG is appealing for expanding recoverable coal reserves, but as with the combustion of all fossil fuels, there are associated greenhouse gas emissions. Coal is the most carbon-intensive of all fossil fuels and has high associated CO_2 emissions [4]. The calorific value of fossil fuel sources varies, with typical values of 50 GJ/tonne for natural gas, 45 GJ/tonne for crude oil, and 30 GJ/tonne for coal [4]. Hence, coal has the highest CO_2 emissions per unit of thermal energy produced [13]. If coal is to become a major contributor in the future global energy supply, CO_2 capture and storage techniques would need to be incorporated in the process. UCG has good potential for CO_2 reduction. During gasification, CO_2 is produced, which can be captured from the syngas and stored for long terms. If UCG is successfully linked to such carbon capture and storage (CCS), a method will be available for exploiting the energy in previously unrecoverable coal reserves while satisfying standards for reducing CO_2 emissions.

The aim of this paper is to review key areas and technologies for underground coal gasification so as to provide insights into this developing coal conversion method.

Underground Coal Gasification

Brief UCG History

The concept of coal gasification has been recognized for more than 200 years and was first used during the late 1800s to produce town gas fuel for heating and lighting applications [14]. Today, coal gasification is primarily used to provide fuel for advanced power plants and chemical feedstocks for use in the chemical industry [9]. Conventionally, coal is extracted from the ground through mining, processed, transported, and then gasified in a surface gasifier unit to produce syngas. Underground coal gasification is a combined extraction and conversion process.

Experimentation on UCG was first performed in 1912 by Sir William Ramsay in England [4]. The experiments demonstrated the potential of UCG, but Ramsay's work was interrupted by the First World War. After the war, further UCG research did not continue, since coal was relatively inexpensive and available through conventional mining techniques in Western Europe [4]. The former Soviet Union was the first to begin considerable research and development programs with respect to large-scale UCG systems in the 1930s [15,16]. UCG technology was first utilized within commercial operations by the former Soviet Union for heating and power generation applications, which has continued to implement these systems for over 50 years [17]. Even though UCG has the appearance of being commercially mature, the technology from the former Soviet Union has been gaining interest only recently, with a rapid increase in the number of pilot plants throughout the rest of the world over the last decade. There are many commercial projects entering pilot plant phase and undergoing study, in Australia, New Zealand, the USA, India, Pakistan, Canada, and Italy. National research programs are being reconsidered in the USA and the UK, after preliminary systems failed to reach commercial maturity. As of 2008, the number of UCG trials includes 200 in the former Soviet Union, 33 in the USA, and approximately 40 distributed between South Africa, China, Australia, Canada, New Zealand, India, Pakistan, and Europe [17,18].

UCG Concepts and Technology

UCG is similar to surface gasification [19], with syngas produced through the same chemical reactions [12]. The main difference is that surface gasification occurs in a manufactured reactor whereas the reactor for a UCG system is a natural geological formation containing unmined coal [19, 20]. UCG also has similarities to *in situ* combustion processes applied in heavy-oil recovery and oil shale retorting, with such common operational parameters as roof/floor stability, seam continuity and permeability, and ground water influx [19, 21].

The basic UCG concept is illustrated in Figure 1. UCG involves an arrangement of injection and production wells drilled into coal seams. The coal is ignited and compressed gasification agents are fed into the coal seam through injection wells which triggers and controls an *in situ* sub-stoichiometric combustion process, producing syngas

[22]. Syngas is extracted using production wells and is processed for use [17]. Suitable gasifying agents include air, oxygen, steam/air, and steam/oxygen [23]. The main difference between using oxygen and atmospheric air is that utilization of oxygen increases the heating value of the syngas [24], but producing pure oxygen requires additional energy and resources.

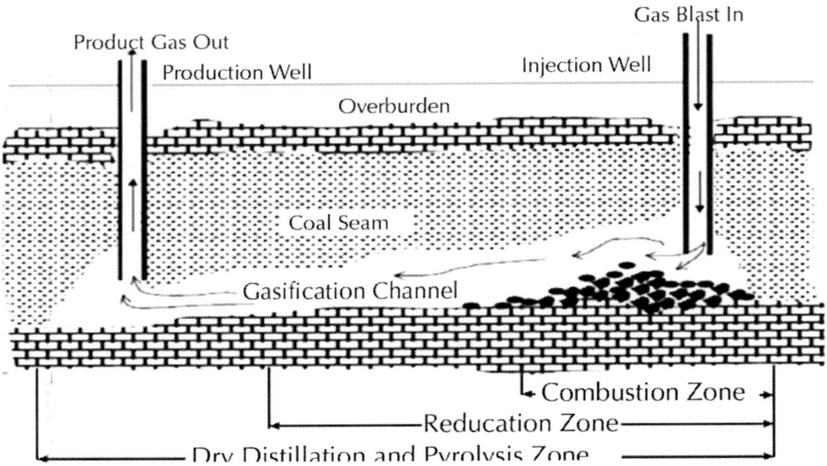

Figure 1: Schematic of *in situ* underground coal gasification process (modified from [21]).

Coal ignition is initiated through the use of an electric coil or gas firing near the face of the coal seam. Continuous oxidant flow through the injection well allows for gasification to be sustained [1]. The temperature of the gasification process is maintained through varying the oxidant flow to the reactor [23]. In UCG systems, the temperature of the coal face can reach temperatures in excess of 1,500 K [24, 25].

Various chemical reactions, temperatures, pressures, and gas compositions exist at different locations within a UCG gasifier. The gasification channel is normally divided into three zones: oxidization, reduction, and dry distillation and pyrolysis [26, 27]. In the oxidization zone, multiphase chemical reactions occur involving the oxygen in the gasification agents and the carbon in the coal the highest temperatures in the gasifier occur in the oxidation zone, due to the large release of energy during the initial reactions [28]. The following reactions occur in the oxidation zone:

$$C+O_2 \rightarrow CO_2+393.8kJ \quad (1)$$

$$2C+O_2 \rightarrow 2CO+231.4kJ \quad (2)$$

$$2CO+O_2 \rightarrow 2CO_2+571.2kJ \quad (3)$$

In the reduction zone, the main reactions involve the reduction of H_2O (g) and CO_2 into H_2 and CO at high temperatures within the oxidation zone [26]. The following endothermic reactions occur in the reduction zone [21, 28]:

$$C+CO_2 \rightarrow 2CO-162.4kJ \quad (4)$$

$$C+H_2O\ (g) \rightarrow CO+H2-131.5kJ \quad (5)$$

Under the catalytic action of coal ash and metallic oxides, a methanation reaction occurs:

$$C+2H_2 \rightarrow CH_4+74.9kJ \quad (6)$$

The energy terms, within the above equations, represent the amount of energy released or consumed during each reaction with the stoichiometric coefficients in equations representing moles. Equations 1 to 6 are taken from [21] and [28]. The endothermic reactions in the reduction zone decrease the temperature in the gasification channel to below that required for the reduction reactions.

Within the distillation (pyrolysis) zone, the coal seam is decomposed into multiple volatiles including H_2O, CO_2, CO, C_2H_6, CH_4, H_2, tar, and char [21,28]. At the exit of the gasification channel, the volatile composition of the syngas consists mostly of CO, H_2, and CH_4. The UCG process can also have other products, including H_2S, As, Hg, Pb, and ash [27, 29, and 30]. The composition of syngas is highly dependent on the gasification agent, air injection method, and coal composition [31, 32]. During operation, the three gasification zones move along the direction of the air flow, ensuring continuous gasification reactions [21]. A distinguishing feature of UCG, compared to surface gasification, is that drying, pyrolysis, and char gasification occur simultaneously within the coal [26].

By-products of the UCG process pose an environmental hazard to the local surroundings through leaching of organic and inorganic materials into groundwater. Environmental data were first made available after later trials in the USA, including Hanna and Hoe Creek UCG trials, for which groundwater contamination monitoring was conducted before, during, and after gasification. The results illustrated that at shallow depths, UCG can pose a significant risk to groundwater in adjacent strata [30].

Groundwater pollution around UCG zones is mainly caused by one of the following mechanisms: dispersion and penetration of the pyrolysis products of the coal seam to the surrounding rock layers, the emission and dispersion of high contaminants with gas products after gasification, and migration of residue by leaching and penetration of groundwater [30]. In addition, the escaped gases such as carbon dioxide, ammonia, and sulfide may change the pH value of the local strata if dissolved.

The entire process is confined to the space of the coal seam and is sealed from the surface by natural geological formations or man-made barriers; the coal seam and strata serve, to some extent, as a natural groundwater cleaning system. In general, systems have active pressure control, in which the cavity pressure is held in equilibrium or below that of the surrounding strata [17, 30]. The pressure difference induces flow into the reactor space, which inhibits gasification products from leaking away from the cavity [33, 34].

The quality of the product gas is influenced by several parameters - such as the pressure inside the coal seam, coal properties, feed conditions, kinetics, and heat and mass transport within the coal seam - and the product of the UCG process is a multi-compound, high-temperature, and high-pressure syngas [1]. When the syngas reaches the surface, it is cleaned and undesired by-products are removed from the product stream [24]. Removal techniques are similar to those used with surface gasifiers. Once the by-products are removed, they can be disposed of safely, or used for other chemical processes [16]. The degree of cleaning required is dependent on the use of the syngas; syngas is cleaned either to meet the specification for input into a gas turbine (for electricity generation) or to be of sufficient purity for use as a chemical feedstock for conversion to synthetic fuels [20].

Methods for UCG

Two standard methods of preparing a coal seam for gasification have been utilized successfully: shaft and shaftless. The method implemented is dependent on parameters such as the natural permeability of the coal seam, geochemistry of the coal, seam thickness, depth, width and inclination, proximity of urban developments, and the amount of mining desired [34].

Shaft UCG methods

Shaft methods use coal mine galleries and shafts to transport gasification reagents and products, which sometimes entail the creation of shafts and the drilling of large-diameter openings through underground labor [34]. The shaft method was the first technique utilized within UCG systems. Currently, the shaft method is only employed in closed coal mines due to economic and safety reasons [34]. The following are examples of common UCG shaft methods:

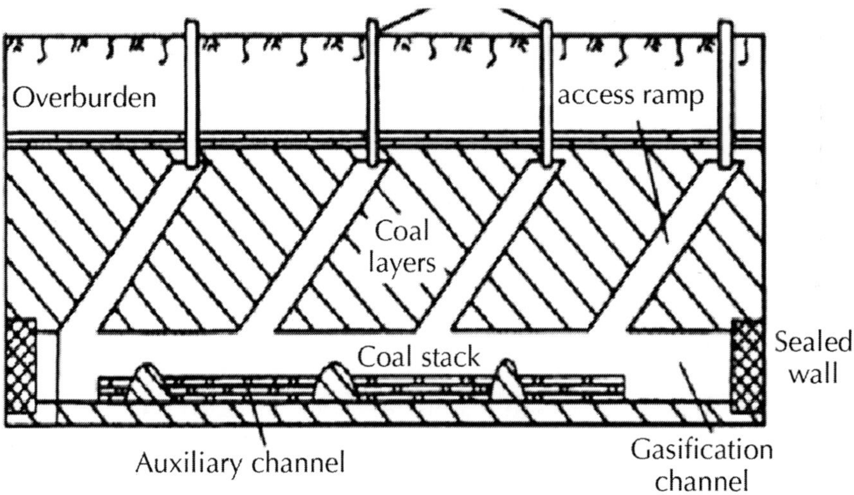

Figure 2: Schematic of the structure of a LLT underground gasifier (modified from [28]).

Chamber or warehouse method

This method utilizes constructed underground galleries with brick walls separating coal panels. Gasification agents are supplied to a previously ignited coal face on one side of the wall, and the syngas is removed from a gallery on the other side. The chamber method strongly relies on the natural permeability of the coal seam to allow for sufficient oxidant flow through the system. The syngas composition may vary during operation, and the gas production rates are often low. To improve system output, coal seams are often outfitted with explosives for rubblization prior to the reaction zone [21].

Borehole producer method

For this method, parallel underground galleries are created within a coal seam with sufficient distance between them. The galleries are connected by drilling boreholes from one gallery to the other [34]. Remote electric ignition of the coal in each borehole is used to initiate the gasification process. This method is designed to gasify considerably flat-lying seams. Some variations exist where linking of the galleries is accomplished through hydraulic and electric linking [21, 34].

Stream method

This method is designed for sharply inclined coal beds. Parallel pitched galleries following the contour of the coal seam are constructed and are connected at the bottom of the seam by a horizontal gallery also known as a fire-drift. To initiate gasification, fire is introduced within the horizontal gallery. The hot coal face moves up the seam slope with oxidant fed through one inclined gallery and syngas leaving through the other. The main advantage of this method is that the ash and roof material drop down to fill the void space created during the process, which prevents suffocating the gasification process at the coal front [21].

LLT gasification method

This method utilizes mined tunnels or constructed roadways to connect the injection well to the production well [4]. Typical long and large tunnel (LLT) systems consist of a gasification channel, two auxiliary holes, and two auxiliary tunnels (Figure 2). The auxiliary holes are arranged between the injection and production wells and are used as malfunction holes for the injection of air and water vapor, or to discharge gas for added gasifier control. LLT also includes an auxiliary tunnel constructed of bricks, which is an auxiliary installation for air injection that prevents blockage in the gasification channel. The mined tunnels are isolated by sealing walls to prevent leakage of combustible gases from the gasifier [35]. The location and height of the oxidant injection points and gas outlet points can be adjusted, allowing for two-dimensional control of oxidant injection and gas production [28].

Shaftless UCG methods

Recently, most of the focus of research has been on the shaftless methods, which employ directional drilling techniques [6]. The preparation of a reactor for the directional drilling technique consists of the creation of dedicated in-seam boreholes for oxidant injection and product collection using drilling and completion technology that has been adapted from oil and gas production.

With shaftless methods, all preparation and operational processes are carried out through a series of boreholes drilled from the surface into a coal seam and do not require underground labor. Preparation of a shaftless reactor consists of the creation of dedicated in-seam boreholes for oxidant injection and product collection using drilling and completion technology that has been adapted from oil and gas production [34]; the approach generally includes drilling inlet and outlet boreholes into a coal seam, increasing the coal permeability between the inlet and outlet boreholes, igniting the coal seam, introducing an oxidant for gasification, and extraction of the product gas from the outlet well [21]. Currently, there are two main classifications of shaftless UCG methods: linked vertical well (LVW) and controlled retractable injection point (CRIP).

LVW method

The LVW method is one of the oldest methods for UCG and is derived from technology developed in the former Soviet Union [16]. Vertical wells are drilled into a coal seam, and internal pathways in the coal are utilized to direct the oxidant and product gas flow from the inlet to the outlet borehole. Internal pathways can be naturally occurring or constructed [35]. In its simplest form, the LVW method has inlet and outlet borehole locations that are static for the life of the system. During operation, the coal face migrates and it is found that system control, performance, and syngas quality are affected negatively as the distance from the coal face to the oxidant injection point increases [4]; this factor greatly reduces the feasibility of simple LVW systems.

A more advanced LVW approach involves a series of dedicated injection boreholes located along the length of a coal seam [21]. Over the life of a UCG reactor, the coal face, being gasified, travels as localized coal is exhausted [4]. Having multiple boreholes for injection allows for improved static operating conditions. A more complex variation of the LVW method also exists where multiple inlet and outlet boreholes are drilled into a coal seam, forming inlet and outlet borehole pairs. Parallel inlet and outlet manifolds are connected to the boreholes to provide a path for oxidant and syngas flows, respectively. Coal between each pair of inlet and outlet boreholes forms a zone. When the coal in a zone has been exhausted, new boreholes are drilled in a location of fresh coal, forming new zones [21].

Low-rank coals, such as lignites, have considerable natural permeability and can be exploited for UCG without the need for linking technologies. However, high-rank coals, such as anthracites, are far less permeable, making the gas production rate more limited if UCG is employed [35]. For the use of high-rank coals in UCG, a method of linking must be employed to increase the permeability and fracture the coal seam [36]. The boreholes in traditional LVW gasifiers are linked by special methods including forward combustion, reverse combustion, fire linkage, electric linkage, hydrofracturing, and directional drilling to create sizable gasification channels [35, 37].

Controlled retractable injection point

Over the span of a coal seam, the geometry may change, resulting in variable UCG operation and system performance [38]. In the past, this problem was solved by having multiple injection and/or production wells so that static operating conditions could be accomplished through moving the gasifier zones to fresh coal [16]. CRIP offers an alternative approach where the vertical injection well is not moved, but the injection point is moved within the coal seam to fresh coal when necessary [39].

The CRIP method relies on a combination of conventional drilling and directional drilling to access the coal seam and physically form a link between the injection and production wells, without the use of linking technologies utilized in LVW methods [38]. A vertical section of injection well is drilled to a predetermined depth, after which directional drilling is used to expand the hole and drill along the bottom of the coal seam creating a horizontal injection well [40]. At the end of the injection well, a gasification cavity is initiated in a horizontal section of the coal seam, creating a localized reactor. The CRIP system utilizes a burner attached to retractable coiled tubing which is used to ignite the coal [39]. The burner burns through the borehole casing to ignite the coal. The ignition point can be moved to any desired location along the horizontal injection well for the creation of a new gasification cavity after a deteriorating reactor has been deserted [38]. Typically, the injection point is retracted using a gas burner, which burns a section of the liner at a desired location [39]. In this manner, accurate control of the gasification process can be obtained. This UCG method has gained popularity in Europe and the USA, but the use of the CRIP method for UCG is fairly new and currently has not become commonly employed [4].

UCG with CO_2 Capture and Storage

All fossil fuels emit CO_2 when combusted. Currently, coal has the highest CO_2 emissions, per unit energy produced, of the fossil fuels used in combustion [13, 41]. To maintain and expand the use of coal, implementation of CCS technologies is becoming imperative.

CO_2 capture can be performed in three main fashions: pre-combustion, post-combustion, and oxy-firing [42]. A broad range of technology options are available for capturing CO_2 including physical absorption, chemical absorption, membrane separation, and cryogenic separation [42, 43]. Within UCG, the syngas compositions, temperatures, and pressures of production streams at the exit of a production well are comparable to those of surface gasifiers, which allow similar methods of CO_2capture. Due to similarities, it is believed that UCG syngas could take advantage of separation using physical sorbents, within a pre-combustion arrangement, which has costs comparable to capture technologies commonly utilized in integrated gasification combined cycles [4, 12]. Post-combustion methods are also applicable and would be directly comparable in terms of cost and performance to typical post-combustion systems utilized in power plants. Oxy-firing options are possible for UCG as well, and within a power generating scenario, an air separation unit can generate O_2 streams for injection into the UCG and for use in an oxy-fired plant utilizing the syngas [12].

The spatial coincidence of geological carbon storage (GCS) options with UCG opportunities suggests that designers could colocate and combine UCG and GCS systems with high potential for effective CO_2 storage [18]. In general, these storage options would be the same for conventional carbon sequestration operations, including saline formations and mature oil and gas fields [44]. For UCG-CCS utilizing conventional sequestration options, there could exist common interests in site characterization and monitoring between UCG and CCS projects, where work performed during the design and implementation of one project could be used within the other. Coordinating UCG and CCS designs would improve economics for both projects.

If UCG and CCS are coupled, there is an attractive carbon management scheme associated, where most of the expected CO_2 emissions are sequestered back into a coal seam void that has been recently created by spent subsurface reactors through existing injection and production wells [13, 21]. When voids are created, they typically collapse, similar to voids produced during longwall coal mining, leaving zones of artificial breccias with high permeability. Suitable containment zones prevent vertical flow of CO_2 to the surface, where storage locations are isolated from the surface by low-permeability strata (known as seals or caprocks, often shales or evaporites) [4,45]. For a spent UCG system to accommodate CO_2 storage, the void must be at depths below approximately 800 to 1,000 m [44-46]. These depths are required so that supercritical pressures and temperatures exist that allow the CO_2 density to be high enough (approximately 500 to 700 kg/m^3) to limit the storage volume required [45].

The UCG-CCS approach, if successful, could offer an integrated energy recovery and CO_2 storage system, which exploits a new sequestration resource created during operation. A significant challenge with CCS is its large energy requirement [47], of which a considerable portion is consumed during CO_2 capture and compression [48]. The pressure after compression is generally high enough to allow for a reduction in pressure during transport while allowing the fluid to be in a liquid state [9]. If CO_2 storage is accommodated in spent UCG reactors, CO_2 transport and compression requirements decline. CO_2 transport accounts for 5% to 15% of a conventional CCS financial budget, which can be lowered with a self-contained UCG-CCS project, through reduced piping and shipping requirements associated with long-distance transport [18]. A large portion of the budget for a CCS project is allotted for CO_2 storage, typically 10% to 30%, most of which is used for geological and geophysical studies and drilling injection wells [18, 48]. These tasks are commonly completed during UCG construction and would not need to be repeated for the implementation of CCS, thus reducing system cost relative to conventional storage methods [18].

As of 2009, it remains unclear if CCS using UCG-produced voids is viable [44]. Until recently, this alternative has received little attention, and there remains substantial scientific uncertainty associated with the technological challenges and environmental risks of storing CO_2 in this manner [13,44]. For full-scale commercialization, extensive research and development is needed to alleviate the uncertainties. Currently, CO_2 sequestration is under development internationally by such organizations as the Intergovernmental Panel on Climate Change and Carbon Sequestration Leadership Forum [13].

CONCLUSIONS

Although the earth is an abundant source of coal, a significant amount is currently unrecoverable. With the introduction of UCG, recoverable coal reserves can be expanded by possibly a couple hundred years. Coal is likely to remain used in many countries, increasing the needs for new technologies that permit more environmentally benign extraction and utilization. Wide-scale use of UCG is such a technology option, with the syngas it produces usable as a fuel. Fossil fuels typically utilized in power production could then be used for other purposes, which would result in large reductions in their consumption rates.

UCG offers a coal extraction and conversion method in a single process that avoids many of the challenges associated with conventional mining practices. UCG has a high potential for integration with CCS using conventional methods utilized in power production due to similarities with surface gasifier units. UCG also has the potential to store CO_2 within voids created during its operation, which reduces the need for transport and storage site identification. In essence, UCG could provide a cost-effective, near-zero-carbon, energy source through the use of a self-contained system with a closed carbon loop.

AUTHORS' CONTRIBUTIONS

SS carried out the review and drafted the manuscript. BR and MR co-supervised the investigation, conceived of the study, reviewed the manuscript, and helped draft parts. All authors read and approved the final manuscript.

ACKNOWLEDGMENTS

The authors acknowledge the financial support of the Natural Sciences and Engineering Research Council of Canada.

REFERENCES

1. Daggupati, S, Mandapati, RN, Mahajani, SM, Ganesh, a, Pal, AK, Sharma, RK, Aghalayam, P: Compartment modeling for flow characterization of underground coal gasification cavity. Ind. Eng. Chem. Res... 50, 277–290 (2011).
2. Tanaka, N(e): World Energy Outlook 2009, International Energy Agency, Paris (2009)
3. Energy Information Administration: International Energy Outlook 2010, Department of Energy, U.S. Government, Washington (2010)

4. Roddy, DJ, Younger, and PL: Underground coal gasification with CCS: a pathway to decarbonising industry. Energy Environ. Sci... 3, 400–407 (2010).
5. Ediger, VS, Hosgor, E, Surmeli, AN, Tatlidil, H: Fossil fuel sustainability index: an application of resource management. Energy Policy. 35, 2969–2977 (2007).
6. Hammond, GP: Energy, environment and sustainable development: a UK perspective. Transactions of the Institution of Chemical Engineers, Part B. 78, 304–323 (2000)
7. Aleklett, K, Hook, M, Jakobsson, K, Lardelli, M, Snowden, S, Soderbergh, B: The peak of the oil age–analyzing the world oil production reference scenario in World Energy Outlook 2008. Energy Policy. 38, 1398–1414 (2010).
8. World Energy, C: Survey of Energy Resources 2007, London, World Energy Council (2007)
9. Ghose, MK, Paul, and B: Underground coal gasification: a neglected option. Int. J. Environ. Stud... 64(6), 777–783 (2007).
10. Couch, G: Underground Coal Gasification, IEA Clean Coal Centre, London (2009)
11. Shackley, S, Mander, S, Reiche, a: Public perceptions of underground coal gasification in the United Kingdom. Energy Policy. 34, 3423–3433 (2006).
12. Burton, E, Friedmann, J, Upadhye, R: Best Practices in Underground Coal Gasification, Lawrence Livermore National Laboratory, Livermore (2006)
13. Khadse, A, Qayyumi, M, Mahajani, S, Aghalayam, P: Underground coal gasification: a new clean coal utilization technique for India. Energy. 32, 2061–2071 (2007).
14. Breault, RW: Gasification processes old and new: a basic review of the major technologies. Energies. 3, 216–240 (2010).
15. Gregg, DW, Hill, RW, Olness, DU: An Overview of the Soviet Effort in Underground Gasification of Coal, Lawrence Livermore National Laboratory, Livermore (1976)
16. Shafirovich, E, Varma, A: Underground coal gasification: a brief review of current status. Ind. Eng. Chem. Res... 48, 7865–7875 (2009).

17. Van der Riet, M: Underground coal gasification. Proceedings of the SAIEE Generation Conference, Eskom College, Midrand (19 Feb 2008)
18. Roddy, D, Gonzalez, G: Underground coal gasification (UCG) with carbon capture and storage (CCS). In: Hester RE, Harrison RM (eds.) Issues in Environmental Science and Technology, pp. 102–125. Royal Society of Chemistry, Cambridge (2010)
19. Pana, C: Review of Underground Coal Gasification with Reference to Alberta's Potential, Energy Resources Conservation Board, Edmonton (2009)
20. Walker, L: Underground coal gasification: a clean coal technology ready for development. The Australian Coal Review. 8, 19–21 (1999)
21. Lee, S, Speight, JG, Loyalka, SK: Handbook of Alternative Fuel Technologies, CRC, Boca Raton (2007)
22. Kempka, T, Plötz, ML, Schlüter, R, Hamann, J, Deowan, SA, Azzam, R: Carbon dioxide utilisation for carbamide production by application of the coupled UCG-urea process. Energy Procedia. 4, 2200–2205 (2011)
23. Perkins, G, Sahajwallaa, and V: Modelling of heat and mass transport phenomena and chemical reaction in underground coal gasification. Chem. Eng. Res. Des.. 85(3), 329–343 (2007).
24. Perkins, G, Sahajwalla, V: Steady-state model for estimating gas production from underground coal gasification. Energy Fuel. 22, 3902–3914 (2008).
25. Peng, FF, Lee, IC, Yang, and RYK: Reactivities of in situ and ex situ coal chars during gasification in steam at 1000-1400 °C. Fuel Process. Technol... 41, 233–251 (1995).
26. Perkins, G, Sahajwallaa, V: A mathematical model for the chemical reaction of a semi-infinite block of coal in underground coal gasification. Energy Fuel. 19, 1679–1692 (2005).
27. Yang, LH, Pang, XL, Liu, SQ, Chen, and F: Temperature and gas pressure features in the temperature-control blasting underground coal gasification. Energy Sources: Part A. 32, 1737–1746 (2010).
28. Yang, L, Liang, J, Yu, and L: Clean coal technology—study on the pilot project experiment of underground coal gasification. Energy. 28, 1445–1460 (2003).

29. Liu, S, Wang, Y, Yu, L, Oakey, J: Thermodynamic equilibrium study of trace element transformation during underground coal gasification. Fuel Process. Technol... 87, 209–215 (2006).
30. Shu-qin, L, Jing-gang, L, Mei, M, Dong-lin, D: Groundwater pollution from underground coal gasification. J. China Univ. Mining & Technol... 17(4), 0467–0472 (2007).
31. Stańczyk, K, Howaniec, N, Smoliński, A, Świadrowski, J, Kapusta, K, Wiatowski, M, Grabowski, J, Rogut, J: Gasification of lignite and hard coal with air and oxygen enriched air in a pilot scale ex situ reactor for underground gasification. Fuel. 90, 1953–1962 (2011).
32. Prabu, V, Jayanti, S: Integration of underground coal gasification with a solid oxide fuel cell system for clean coal utilization. Int. Journal of Hydrogen Energy. 37, 1677–1688 (2012).
33. Yang, LH: A review of the factors influencing the physicochemical characteristics of underground coal gasification. Energy Sources, Part A. 30(11), 1038–1049 (2008).
34. Wiatowski, M, Stańczyk, K, Świądrowski, J, Kapusta, K, Cybulski, K, Krause, E, Grabowski, J, Rogut, J, Howaniec, N, Smoliński, A: Semi-technical underground coal gasification (UCG) using the shaft method in Experimental Mine "Barbara". Fuel. 99, 170–179 (2012)
35. Liang, J, Liu, S, Yu, L: Trial study on underground coal gasification of abandoned coal resource. In: Xie H, Golosinki TS (eds.) Proceedings of the ‹99 International Symposium on Mining Science and Technology, Beijing, August 1999. Mining Science and Technology 99, pp. 271–275. A.A. Balkema, Rotterdam (1999).
36. Blinderman, MS, Klimenko, AY: Theory of reverse combustion linking. Combustion and Flame. 150, 232–245 (2007).
37. Blinderman, MS, Saulov, DN, Klimenko, AY: Forward and reverse combustion linking in underground coal gasification. Energy. 33, 446–454 (2008).
38. Nourozieh, H, Kariznovi, M, Chen, Z, Abedi, J: Simulation study of underground coal gasification in Alberta reservoirs: geological structure and process modeling. Energy Fuel.24, 3540–3550 (2010).

39. Klimenko, AY: Early ideas in underground coal gasification and their evolution. Energies. 2, 456–476 (2009).
40. Wang, GX, Wang, ZT, Feng, B, Rudolph, V, Jiao, and JL: Semi-industrial tests on enhanced underground coal gasification at Zhong-Liang-Shan coal mine. Asia-Pac. J. Chem. Eng... 4, 771–779 (2009).
41. Nag, B, Parikh, J: Indicators of carbon emission intensity from commercial energy use in India. Energy Economics. 22, 441–461 (2000).
42. Göttlicher, G, Pruschek, R: Comparison of CO2 removal systems for fossil-fuelled power plant processes. Energy Convers. Mgmt... 38, 173–178 (1997).
43. Ho, MT, Allinson, G, Wiley, DE: Comparison of CO_2 separation options for geo-sequestration: are membranes competitive?. Desalination. 192, 288–295 (2006).
44. Friedmann, SJ, Upadhye, R, Kong, FM: Prospects for underground coal gasification in carbon-constrained world. Energy Procedia. 1(1), 4551–4557 (2009).
45. Orr, FM: Onshore geologic storage of CO_2. Science. 325(5948), 1656–1658 (2009).
46. Budzianowski, WM: Value-added carbon management technologies for low CO2 intensive carbon-based energy vectors. Energy. 41, 280–297 (2012).
47. Steinberg, M: Fossil fuel decarbonization technology for mitigating global warming. International Journal of Hydrogen Energy. 24, 771–777 (1999).
48. Gibbins, J, Chalmers, H: Carbon capture and storage. Energy Policy. 36, 4317–4322 (2008).

Chapter 4

Chemical and Spectroscopic Characterization of Insoluble and Soluble Humic Acid Fractions at Different pH Values

Andrea Baglieri[1], Daniela Vindrola[2], Mara Gennari[1], and Michèle Negre[2]

[1]Dipartimento di Scienze delle Produzioni Agrarie e Alimentari, University of Catania, Via S. Sofia 98, Catania (CT), 95123, Italy

[2]Dipartimento di Scienze Agrarie, Forestali e Alimentari, University of Turin, Largo Paolo Braccini 2, Grugliasco (TO), 10095, Italy

ABSTRACT

Background

Humic acids (HA) are organic molecules with complex structure and function and variable properties. They are insoluble in strongly acid pH conditions. At present, it is not clear how much is the amount of HA in solution at the pH of natural soils nor are known the characteristics of the different soluble fractions and their possible association with the inorganic phase of soil. The scope of this work was to characterize the soluble and insoluble fractions obtained by acidifying Na humate solution at pH values 3, 5, and 7 and to compare these fractions with the HA obtained at pH 1. At each pH, the precipitate and the soluble fractions were separated and characterized by elemental analysis, total acidity and carboxylic group content, infrared, and ^{13}C NMR and ^{1}H NMR spectroscopy.

Results

The HA fraction insoluble at pH 1 had a high acid group content and aromaticity but a low content of O- and N-containing groups. At pH 3, a fraction with nearly the same characteristics was obtained. At pH 1 and pH 3, the inorganic phase bound to the insoluble humic material was largely constituted by clay minerals and some Al and Fe hydroxides. The soluble fractions at pH values 1 and 3 were very poor and they were composed of a silica gel-like phase associated with polar organic material rich in carboxylic and metal-carboxylate groups. At pH values 5 and 7, only a small fraction of the Na humates precipitated. The fractions remaining in the solution were mainly composed of organic material particularly rich in aromatic and aliphatic groups, while the inorganic phase contained phyllosilicates. The fractions insoluble at pH values 5 and 7 contained a large amount of inorganic material that consisted mainly of phyllosilicates.

Conclusions

The soluble fractions obtained at pH values 5 and 7 represent the humic component that in environmental situations would be dissolved in the soil solution. Our findings could be very useful for a more detailed investigation into the way HA influence plant metabolism under environmental-like conditions, both as regards pH conditions and interaction with the mineral fraction.

BACKGROUND

Soil organic matter can be divided into two major components: organic substances of an individual nature and humic substances (HS). Traditionally, the soil HS have been subdivided into three fractions: fulvic acids, humic acids (HA), and humin.

It is widely recognized that HS are important for the sustainability of terrestrial life. The heterogeneous chemical nature of HS exerts a multifunctional role in the environment by controlling the biogeochemical carbon cycle [1],[2], providing nutrients and bio-stimulants for plant growth [3],[4], and interacting with inorganic [5] and organic pollutants [6],[7].

It is common knowledge that the characteristics of HS depend on the extraction procedure adopted. This fact has been attributed to the complex nature of soil, to the wide range of physical and chemical properties inherent in humic material, and in particular to the intimate association of organic and inorganic soil constituents [8]. Most of the studies reporting the effect of extraction procedures on the chemical structure of HS have concerned the nature of the extraction solvent. These studies pointed out that both the type of soil and the extraction pH played a role in the chemical structure and molecular weight of the humic materials. The procedure recommended by the International Humic Substances Society (IHSS) for preparing HA [9] includes 0.1 M sodium hydroxide extraction followed by 0.1 M hydrochloric acid precipitation. In accordance with the new concept of supramolecular association, the HA obtained by this procedure is composed of many relatively small, chemically diverse organic molecules which form clusters linked by hydrogen bonds (H-bonds) and hydrophobic interactions [10]. A corollary to this model is the concept of micellar structure, i.e., an arrangement of organic molecules in aqueous solution which form hydrophilic exterior regions that shield the hydrophobic interiors from contact with vicinal water molecules [11].

It is assumed that humic acids are completely soluble at pH 13 and completely insoluble at pH 1, but the solubility of HA at intermediate pH values has not been fully investigated. Brigante et al.[12] measured the dissolution kinetics of HA prepared according to the IHSS procedure at different pH values between 10 and 4. HA were completely soluble within less than an hour at pH > 10. At pH between 10 and 9, the dissolution was slower but nearly complete. At pH 8.5 and 8, more than 60% and 40% of the HA respectively were soluble. The separation of HA in a soluble and an insoluble phase at these pH values attests for the heterogeneity of the material defined as humic acid on the basis of its insolubility at pH 1. At our knowledge, the chemical differences between the soluble and insoluble fractions at a given pH have not been investigated.

A drawback of the IHSS procedure is that the HA may contain quite large amounts (e.g., 30%) of inorganic components. Although some attempts have been made to destroy the organo-mineral associations before HS extraction [13],[14], a purification procedure aimed at obtaining ash-free organic fractions is commonly carried out on the extracted humic materials.

Such purification procedure, based on hydrochloric-hydrofluoric acid solubilization of the inorganic fraction, generally decreases the ash content to less than 2%. This is considered a good result as far as the purity of the humic material is concerned. However, little attention has been given to the nature of the inorganic fraction brought in solution by the alkaline extraction and so strongly bound to the humic material to require a drastic treatment to be released.

In this study, a Na humate solution was prepared following the IHSS procedure but without treatment to eliminate the inorganic phase. Aliquots of the humate solution were brought to pH 7, 5, 3, and 1, and the corresponding soluble and insoluble fractions were collected and characterized. The scope was to determine the chemical structure of the organic and inorganic materials precipitated or left in solution at each pH and to obtain some elucidation on the nature of their interactions.

EXPERIMENTAL

Materials

Soil

The HA was extracted from a histosol (Trana, Turin, Italy), studied in another article [15], having the following properties: pH, 6.9; clay, 5.7%; loam, 30.5%; sand, 63.8%; organic carbon, 8.51%; cation exchange capacity, 52.9 meq/100 g; Na oxalate extractable Fe, 0.68%; Na dithionite-citrate-bicarbonate extractable Fe, 0.86%; and Na pyrophosphate extractable Fe, 0.22%.

METHODS

Preparation of the HA Fractions

The soil (100 g subsamples in 500 ml PE flasks) was equilibrated to a pH value of between 1 and 2 with 1 M HCl; the suspension was brought to 500 ml with 0.1 M HCl, shaken for 1 h, centrifuged, and the supernatant eliminated. The solid phase was neutralized to pH 7 with 1 M NaOH, then 0.1 M NaOH was added to provide a final volume of 500 ml. The suspension was shaken under N_2 for 20 h, allowed to settle overnight, then centrifuged (1,500 *g*, 30 min) to separate the supernatant. A series of three alternate treatments with 0.1 M HCl and 0.1 M NaOH was performed to solubilize and remove the fulvic acids. After the final treatment, the precipitate resulting from acidification was dissolved in 0.1 M NaOH, and the whole was treated as illustrated in Figure 1. Equal parts of Na humate were acidified with 0.1 HCl M at different pH values, 1, 3, 5, and 7. At each pH, the soluble and insoluble fractions were separated by centrifugation. Subsequently, the fractions were dialyzed (MW cut-off 3,500 Da) and then freeze-dried.

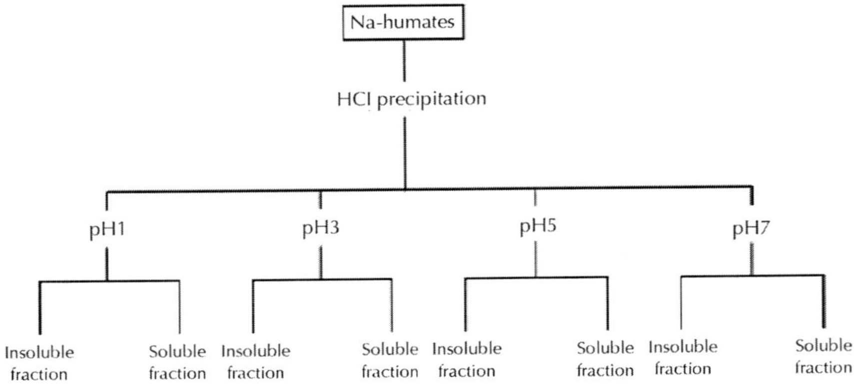

Figure 1: Preparation of insoluble and soluble humic acid fractions at different pH values.

Characterization of the HA Fractions

The HA fractions were characterized for total acidity and carboxylic group content in accordance with the procedure proposed by Swift [9]; measurement of the phenol OH was obtained by difference; the C, H, N content was determined using a C, H, and N LECO Analyzer model 600 (LECO Corp., St. Joseph, MI, USA); O content was calculated by difference. Ash content was determined by incineration at 650°C for 4 h.

The FT-IR spectra were obtained with a Perkin-Elmer FT-IR 2000 spectrometer (Perkin Elmer Italia, S.p.A, Milan, Italy), equipped with an IR source, KBr beam splitter and DTGS KBr detector. For each sample, 64 scans were recorded with a 4-cm^{-1} resolution over a 4,000 to 400 cm^{-1}range on pellets obtained by pressing a mixture of 1 mg organic sample and 400 mg dried KBr.

Solid-state ^{13}C MAS NMR spectra were recorded, fully proton-decoupled, on a Bruker MSL 200 NMR (Bruker Corp., The Woodlands, TX, USA) instrument operating at 50.3 MHz. Rotors with a 7-mm diameter were filled with about 50 mg of the sample; the spinning rate was 4,000 Hz s^{-1}. The following experimental parameters were adopted: spectral width 20,000 Hz, data points 2 K, 100,000 scans, 4 μs, 90 degrees of excitation pulse, and 4 s of relaxation delay. The HPDEC pulse sequence was used with a decoupling power of 300 W

(9H). The FID were zero-filled and processed with 5 Hz line broadening. Chemical shifts refer to ^{13}CO of glycine, set at 176.03 ppm.

The ^{1}H NMR spectra were recorded using a JEOL EX-400 spectrometer (JEOL, Peabody, MA, USA) equipped with a 5 mL probe operating at 399.78 MHz. Samples were prepared by dissolving 20 mg of the sample in 0.6 mL of 0.1 M NaOD/D_2O. After filtration through a glass wool filter, the samples were placed in NMR tubes for analysis. All the samples were adjusted to pH 12. The ^{1}H NMR spectra were recorded using the presaturation sequence for selective saturation of the HDO resonance [16]. The main NMR parameters were 45° pulse width, pre-irradiation 10 s, acquisition time 5.46 s, and 64 scans for a total acquisition time of about 16 min.

The iron and aluminum oxide contents were evaluated by Na dithionite-citrate-bicarbonate extraction [17] at 25°C (Fe DCB and Al DCB, respectively). Aqua regia was used to attack the metals, and then the total iron and aluminum in the humic fractions were analyzed.

All chemicals were of analytical grade, and three replicates per analytical determination were carried out.

Analytical data were evaluated by one-way ANOVA ($P < 0.005$) followed by the Tukey's test for multiple comparison procedures.

RESULTS

Yield and Ash Content

The yield, ash content, and elemental composition of the humic fractions at each pH are reported in Table 1.

Table 1: Yield, ash content, and elemental composition of the humic fractions

Fraction		Yield (g kg−1soil)	Asha(g kg−1humic fraction)	Elemental compositionb(g kg−1humic fraction)				Molar ratio
				C	H	N	O	O/C
pH 1	Insoluble	83.9	247	541a	42a	43a	374a	0.5
	Soluble	2.3	196	478bd	44a	43a	434bc	0.7

pH 3	Insoluble	88.0	267	525a	45a	42a	388ab	0.5
	Soluble	7.2	127	493bc	41a	45a	421abc	0.6
pH 5	Insoluble	22.2	542	478bd	55b	40a	427bc	0.7
	Soluble	60.3	225	518ac	49b	41a	392ab	0.6
pH 7	Insoluble	7.4	609	451d	52b	39a	459c	0.8
	Soluble	67.3	283	534a	49b	42a	375a	0.5

[a]Values are the means of three replicates, and the standard error values are < 10%; [b]on a moisture and ash-free basis. Within columns, values followed by the same letter are not significantly different (P < 0.05).

Baglieri et al.

Baglieri et al. Chemical and Biological Technologies in Agriculture 2014 1:9 doi: 10.1186/s40538-014-0009-x

As expected, the yield of the insoluble fractions decreased as pH increased, while the contrary occurred for the soluble fractions. The total humic material (soluble + insoluble) extracted at each pH was not about the same (86.2, 95.2, 82.5, and 74.7 g kg^{-1} soil at pH 1, 3, 5, and 7, respectively). The observed differences, in particular the low total recovery at pH 7, suggest some loss of material, probably inorganic and organic molecules smaller than the dialysis membrane MW cut-off.

The sum of the ash content of the soluble and insoluble fractions is not the same at each pH. However, this difference is only apparent when considering the value corrected by the yield (21.2, 24.5, 25.6, and 23.6 g at pH 1, 3, 5, and 7 respectively, for extracts from 1 kg of soil). This is demonstrated by the following calculation:

$$\text{Gram of ash at each pH} = \left(\text{Yeld IF} \times \frac{\%\text{ IF ash}}{100}\right) + \left(\text{Yeld SF} \times \frac{\%\text{ SF ash}}{100}\right) \tag{1}$$

Where IF is the insoluble fraction and SF is the soluble fraction.

The high ash content in all fractions emphasizes the importance of employing a purification step if ash-free humic fractions are to be obtained. The effectiveness of the purification step in HA prepared from a histosol was demonstrated by Piccolo [18] who reported a decrease in the ash content from 500 to 10 g kg^{-1} following HF-HCl washing. The ash contents of more than 500 g kg^{-1} in the insoluble fractions at pH 5 and pH 7 attest that there is more inorganic than organic material and is the likely reason for the lack of solubility of these fractions.

Elemental Composition

The elemental composition of the humic fractions is reported in Table 1 on a moisture- and ash-free basis. Therefore, the following considerations concern the organic component of each fraction. While the insoluble fractions at pH 1 and 3 were richer in C and poorer in O than the soluble fractions, the O increased at pH 5 and 7 indicating that the precipitation pH of the HA affected the distribution of the functional groups between the precipitated and the liquid phase. The C content was lower at pH 5 and 7 than at pH 1 and 3 in the insoluble fractions, while the opposite was true for the soluble fractions. The O content and O/C ratio were lower in the insoluble fractions at pH 1 and 3 than at pH 5 and 7 indicating that polarity increased along with pH, while the opposite occurred in the soluble fractions. In the case of the N content, no clear trend could be identified indicating that pH did not affect the distribution of the N-containing moieties in the solid or liquid phase. The H content was higher at pH 5 and pH 7 than at pH 1 and pH 3 in both the insoluble and soluble fractions. This would appear to be an artifact, since the sum of each element present in the soluble and insoluble fractions deriving from the same alkali solution should be the same at any pH. The highest H contents were found in the fractions exhibiting the highest ash contents. This suggests that the presence of inorganic components could induce an overestimation of the H content. One explanation for this result is that the elemental analysis includes a high-temperature incineration step (up to 1,000°C) which could provoke the release of some hydration water and lead to the partial decomposition of some inorganic components, such as kaolinite, thereby releasing OH groups and leading to an overestimation of the hydrogen content.

Acidic Groups

The carboxylic, phenolic, and total acidity distribution in each fraction are reported in Table 2. The wet method used to determine the carboxylic group content is based on titration with Ca acetate. The aim of this technique is to convert the complex mixture of acids in HS into an equivalent amount of acetic acid by the reaction [19]:

$$2RCOOH + Ca(CH_3COO)_2 \leftrightarrow Ca(RCOO)_2 + CH_3COOH \quad (2)$$

Table 2: Acidic groups' concentration and distribution of C intensity of13C NMR spectra of humic fractions

Fraction		Acidic groupsa(meq g–1)			13C NMR distribution (%)				HB/ HI
		Carboxyl	Phenol	Total acidity	0 to 45 ppm	45 to 95 ppm	95 to 160 ppm	160 to 195 ppm	
pH 1	Insoluble	3.8	7.3	11.1	23.0	16.0	33.1	28.0	1.3
	Soluble	4.0	6.7	10.7	14.2	58.8	2.9	24.2	0.2
pH 3	Insoluble	3.1	5.9	8.9	21.0	18.3	33.9	26.9	1.2
	Soluble	2.8	6.7	9.4	14.5	42.1	7.5	35.9	0.3
pH 5	Insoluble	ndb	ndb		ndc	ndc	ndc	ndc	-
	Soluble	ndb	ndb		20.8	6.0	38.7	34.5	1.5
pH 7	Insoluble	ndb	ndb		ndc	ndc	ndc	ndc	-
	Soluble	ndb	ndb		19.1	1.3	35.2	44.3	1.2

[a]Values are the means of three replicates. The standard error values are < 10%; [b]analytical method not applicable at these pH; [c]not determinable because of the poor signal-to-noise ratio of the spectra.

Baglieri *et al.*

Baglieri *et al.* *Chemical and Biological Technologies in Agriculture* 2014 **1**:9 doi: 10.1186/s40538-014-0009-x

This method is reliable only if the carboxylic groups are completely combined. This is why the carboxylic group content, measured using the wet chemical method, decreased at decreasing pH. The total acidity was not measurable at pH 5 and 7. This was most likely due to the carbonation of the barium hydroxide solution used in titration, as reported by Stevenson [20]. Consequently, the content of phenolic groups could not be calculated.

^{13}C NMR

The distribution of the different forms of carbon calculated according to the area of the different NMR spectra regions (alkyl, 0 to 45 ppm; N- and O-alkyl, 45 to 95 ppm; aromatic, 95 to 160 ppm; carboxyl, 160 to 195 ppm) are reported in Table 2.

The amount of carboxyl carbon calculated from the integration of the signal in the 160 to 195 ppm region was much higher than expected considering the chemical titration. This discrepancy has already been reported and discussed in the previous papers [21],[22].

The insoluble fractions at pH values 1 and 3 are largely aromatic (33.1% and 33.9% of the total carbon respectively) with a medium amount of aliphatic carbon (23.0% and 21.0% of the total carbon respectively) and a low content of substituted aliphatic groups (16.0% and 18.3% of the total carbon respectively). Similar concentrations of aromatic carbon were found in HA from histosols by ^{13}C NMR investigation [15],[23],[24] and reflect the high maturity of the HA expected in such soils. By contrast, the soluble fractions at pH 1 and pH 3 contain a very low amount of aromatic C (2.9% and 7.5% of the total carbon) but are very rich in aliphatic substituted carbons (58.8% and 42.1% of the total carbon). These results, together with the high O content of these fractions (Table 1), suggest the presence of polar aliphatic compounds (sugars, amino acids, proteins, carboxylic acids) which are most likely responsible for the solubility at low pH values.

Similar results are obtained evaluating the degree of hydrophobicity (hydrophobic/hydrophilic carbons, HB/HI) as proposed by several authors [25]-[28]. This is given by the areas relative to resonance intervals, as follows:

$$\frac{HB}{HI} = \frac{[(0-45) + (95-160)]}{[(45-95) + (160-195)]} \tag{3}$$

The fractions insoluble at pH 1 and 3 have a degree of hydrophobicity (Table 2) greater than the soluble fractions, with values (>1.0) that are consistent with those observed for HA (between 0.61 and 2.87) by Piccolo et al. [26]. On the other hand, the values of the soluble fractions at pH 1 and 3 of HB/HI are more similar to the same authors' findings for fulvic acids (<0.5).

The ^{13}C NMR spectra of the insoluble fractions at pH 5 and 7 with a high ash content exhibit a low signal-to-noise ratio probably due to the presence of paramagnetic compounds as already observed in soils [29] and humic samples [30]. Consequently, no reliable evaluation of the carbon distribution could be made.

The soluble fractions at pH 5 and pH 7 exhibited a high content of aromatic carbon (38.7% and 35.2% of the total carbon respectively) and a low content of N- and O-substituted aliphatic groups. This is in agreement with the decrease in soluble fraction polarity as pH increases, which is assumed on the basis of the elemental analysis and the degree of hydrophobicity.

The organic component of the fractions obtained at pH 5 and 7 was also studied with ^{1}H NMR in order to obtain more information regarding their chemical characteristics. This technique proved easier to use with samples having a higher ash content than did ^{13}C NMR. The ^{1}H NMR data agree well with the data obtained from the ^{13}C NMR but tend to underestimate the aromaticity, above all when the aromatic molecules are substituted with atoms other than hydrogen [31].

The distribution of the different forms of hydrogen was calculated by integration in three regions of the spectrum (0.5 to 3 ppm, 3 to 4.2 ppm, and 6 to 8 ppm) and expressed as a percentage of the total area of each spectrum. The allocation of the three regions was based on the work of Francioso et al. [32]. The region from 0.5 to 3 ppm corresponds to the H forms linked to aliphatic groups, the region from 3 to 4.2 ppm is assigned to the H forms linked to carbohydrate structures, and finally the region between 6 and 8 ppm is allocated to the H forms linked to aromatic molecules.

From Table 3, which relates to the distribution of the ^{1}H forms, it can be seen that the aliphatic region is similar for all the fractions obtained at pH 5 and 7, while the region related to carbohydrates (between 3 and 4.2 ppm) is higher in insoluble fractions than soluble. Moreover, the latter is more aromatic than the insoluble one. In agreement with the elemental analysis, these components are more polar than the soluble fractions (Table 1).

Table 3: Distribution of H intensity of the1H NMR spectra of the humic fractions

Fraction		1H NMR distribution (%)		
		0.5 to 3.0 ppm	3.0 to 4.2 ppm	6.0 to 8.0 ppm
pH 5	Insoluble	45.6	36.7	17.7
	Soluble	45.2	23.2	31.6
pH 7	Insoluble	47.6	40.3	12.1
	Soluble	46.9	20.5	32.6

Baglieri et al.

Baglieri et al. Chemical and Biological Technologies in Agriculture 2014 1:9 doi:10.1186/s40538-014-0009-x

FT-IR

The FT-IR spectra indicate some large differences between the humic fractions, attesting to the different structures of both the inorganic and organic material (Figure 2). The spectra of the insoluble fractions at pH 5 and 7 (containing more than 50% inorganic component) were identical and dominated by the intense absorption of the Si-O vibration centered at 1,030 cm^{-1} typical of montmorillonite [33]. The other bands attributable to the inorganic components have the same frequencies as those of montmorillonite defiled with kaolinite [33],[34] as follows: 3,696 cm^{-1}, OH stretching of inner-surface hydroxyl groups of kaolinite; 3,621 cm^{-1}, OH stretching of hydroxyl groups; 915 cm^{-1}, AlAlOH deformation; 754 cm^{-1}, Si-O perpendicular (kaolinite); 470 cm^{-1}, Si-O-Si deformation; 430 cm^{-1}, Si-O deformation (kaolinite). The band at 529 cm^{-1} could be the Al-O-Si deformation of montmorillonite, although it is generally reported at lower frequencies (521 to 523 cm^{-1}). Some of the typical bands of montmorillonite are missing (790, 623, and 578 cm^{-1}). The lack of such bands has already been observed in soil montmorillonites and attributed to a larger disorder which provokes the broadening and vanishing of these bands [33]. Besides the bands of the inorganic compounds, the main organic bands are the aliphatic absorptions at 2,928 and 2,853 cm^{-1} (CH_2 stretching). Also, the two bands centered at 1,615 and 1,385 cm^{-1}, typical of the asymmetric and symmetric vibrations of carboxylates, reflect the presence of metal carboxylates.

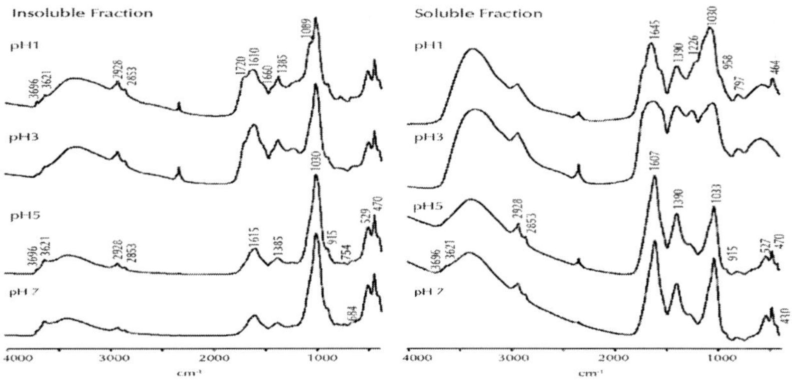

Figure 2: FT-IR spectra of insoluble and soluble humic acid fractions at different pH values.

The spectra of the soluble fractions at pH 5 and pH 7 show the same bands as those attributed to the vibrations of kaolinite and montmorillonite for the insoluble fractions (3,696, 3,620, 1,033, 915, 527, 470, and 430 cm^{-1}) but with lower intensity due to their lower ash content. The COOH vibration is missing, while very strong bands are present at 1,607 ($_{as}COO^-$) and 1,390 cm^{-1}($_{s}COO^-$). The frequency of the $_{as}COO^-$ has been seen to decrease in parallel to an increase in the ionic radii of the metallic ion [35]. According to these authors, the $_{as}COO^-$ band shift toward higher frequencies corresponds to the transition from ionic to covalent bond and can be used to establish the ionic-covalent nature of metal carboxylates. This statement was contested by Piccolo and Stevenson [36], given the possibility that the metal-complexed ketonic group could give a band in the 1,620 to

1,610 cm^{-1} region. However, the $_{as}COO^-$ on the spectra of the soluble fractions at pH 5 and 7 (1,607 cm^{-1}) does not fall in the ketonic region and is between those reported by Vinkler et al. [35] for Na humate (1,590 cm^{-1}) and by Piccolo and Stevenson [36] for K humate at pH 5 (1,610 cm^{-1}). This suggests the formation of ionic humates, probably with a great presence of Na ions in the solution at these pH values.

The spectra of the insoluble fraction at pH 1 and pH 3 exhibit the same, but weaker, bands as those attributed to the inorganic component in the insoluble fractions at pH 5 and 7. In contrast, the HA bands are stronger and some new peaks are present, in particular, the COOH vibration at 1,720 cm^{-1}; the band in the 1,610 to 1,660 cm^{-1} region is broader with a shoulder at 1,660 cm^{-1}and is likely due to an overlap between $_{as}COO^-$ vibration, aromatic -C = C- vibration and N-containing groups. A partial influence on this band can also come from the OH stretching of water bonded to the clay minerals which has been seen to absorb in the 1,630 to 1,660 cm^{-1} range [33].

The spectrum of the insoluble fraction at pH 1 exhibits a strong band at 1,089 cm^{-1} similar to that shown on the spectrum of a montmorillonite and attributed to the Si-O stretching of a cristobalite-like phase impurity [34].

The spectra of the soluble fractions at pH 1 and 3, having the lowest ash content, exhibit the typical features of the HS spectra eliminating part of the diagnostic bands of the inorganic components. The silicates O-Si-O vibration band centered at 1,030 cm^{-1} on the other spectra is broad and shifted at higher frequencies (up to 1,075 cm^{-1} at pH 1). A Si-O vibration at 1,075 cm^{-1}, besides weaker bands at 1,200, 958, 797, and 464 cm^{-1}, has been attributed to the silica gel[37]. The same bands are present on the spectrum of the soluble fraction at pH 1 which also exhibits the two carboxylate group bands (1,645 and 1,390 cm^{-1}). Very similar spectra were reported for the soluble products formed between hydroxy-Al ions and orthosilicic acid in the presence of citric acid [37] and HS [38]. This suggests that the fractions which do not precipitate at low pH are organo-mineral colloids in which the inorganic phase is similar to the silica gel. However, the frequency of the broad band centered at 1,645 cm^{-1} is higher than those reported by Vinkler et al. [35] also in the case of covalent humates with Fe and Al (1,625 cm^{-1}) and should be overlapped by other absorptions, for example, the OH vibration of water molecules bonded to silica gel which occurs at 1,630 cm^{-1}[33].

Although the FT-IR spectra of these fractions provided evidence of the presence of phyllosilicates and silica gel, we cannot exclude that the inorganic fractions also contain iron and aluminum oxides. The FeOH bending and the Fe-O stretching of a Fe(III) synthetic polycation absorb at 550 and 680 cm^{-1} respectively [39],[40]. Other authors report that the main absorption of ferrihydrite took place at 587 cm^{-1} while hematite gave sharp bands at 460 and 580 cm^{-1}[41]. On the spectra for those insoluble fractions at pH 5 and pH 7, a weak band is present at 684 cm^{-1} which could be attributed to Fe-O stretching, while the other bands typical of Fe vibrations could be overlapped or dwarfed by the strong absorption of silicates. The hydroxides are generally identified by several diagnostic bands in the OH stretching region, their number and frequency depending on the grade of crystallinity of the minerals [42]. However, in the presence of silicates with water in the interlayer, these bands have been seen to mix in a single broad absorption band in the 3,000 to 3,630 cm^{-1} range [42]. Therefore, the lack of such bands in our samples does not mean that they do not contain aluminum hydroxides. The other typical Al hydroxide bands are at 1,025, 970, and 531 cm^{-1}[43]. The first two bands are overlapped by Si-O absorption at 1,030 cm^{-1}, but a strong band at 530 cm^{-1} cannot be attributed to silicates and could indicate the presence of aluminum hydroxides.

Metal Concentrations in the Inorganic Components

The total metal content and the DCB extractable concentration of Al and Fe in the humic fractions are reported in Tables 4 and 5. The highest concentrations of Si and Al (as total Al) were found in the insoluble fractions, which indicate the presence of phyllosilicates (Table 4). In the insoluble fractions at pH 5 and pH 7, less than 5% Al was DCB extractable showing that this metal was mainly present in clay mineral structures. In the insoluble fractions at pH 1 and pH 3, 28% and 32% of the total aluminum were DCB extractable indicating that part of the Al was present as oxyhydroxides or HA bound. In contrast, the DCB extractable Fe was more abundant in the insoluble fractions at pH 1 and 3 (64%) than at pH 5 and 7 (34 and 46% respectively). The high proportion of DCB extractable Fe in the HA-rich fractions at pH 1 and

3 suggests the presence of organo-metallic compounds in which the Fe atom could act as a bridge between the humic and inorganic phases.

Table 4: Total and DCB extractable Al and Fe in the humic fractions

Fraction Total (mol kg−1)		Si	Fe			Al		
		Total	DCB extractable		Total	DCB extractable		
		(mol kg−1)	(mol kg−1)	(%)a	(mol kg−1)	(mol kg−1)	(%)a	
pH 1	Insoluble	1.81a	0.47abe	0.30ab	64	1.02a	0.29ac	28
	Soluble	nd	nd	nd	nd	nd	nd	nd
pH 3	Insoluble	1.67ae	0.60ace	0.35ab	58	1.03a	0.33a	32
	Soluble	0.87be	0.24bde	0.24a	100	0.08b	0.10b	125
pH 5	Insoluble	3.67c	0.73c	0.25a	34	2.33cd	0.09b	4
	Soluble	0.84be	0.37de	0.28ab	76	0.46a	0.22c	48
pH 7	Insoluble	4.22d	0.83c	0.37b	46	2.74d	0.10b	4
	Soluble	1.26e	0.38e	0.27ab	71	0.60a	0.26c	43

[a]Percentage of the total. Within columns, values followed by the same letter are not significantly different ($P < 0.05$). nd, not determined.

Baglieri et al.

Baglieri et al. Chemical and Biological Technologies in Agriculture 2014 1:9 doi:10.1186/s40538-014-0009-x

Table 5: Total metal concentration of the humic acid fractions

Fraction (mol kg−1)		Ca	K	Mg	Na
		(mol kg−1)	(mol kg−1)	(mol kg−1)	
pH 1	Insoluble	0.07ad	0.13ad	0.17ad	0.31a
	Soluble	nd	nd	nd	nd
pH 3	Insoluble	0.11abcd	0.18a	0.22a	0.50a
	Soluble	0.14abcd	0.06b	0.03b	1.69b
pH 5	Insoluble	0.12abcd	0.28c	0.55c	0.97c
	Soluble	0.16bcd	0.07b	0.11d	2.32d
pH 7	Insoluble	0.09cd	0.33c	0.67e	1.00c
	Soluble	0.14d	0.10bd	0.17ad	2.21d

Within columns, values followed by the same letter are not significantly different ($P < 0.05$). nd, not determined because of the low amount of sample.

Baglieri et al.

Baglieri et al. Chemical and Biological Technologies in Agriculture 2014 1:9 doi: 10.1186/s40538-014-0009-x

In the soluble fraction at pH 3, all the Al and Fe were DCB extractable attesting to a lack of clay minerals, which agrees with the silica-gel structure deduced from the FT-IR spectra. In the soluble fractions at pH 5 and 7, more than 70% of the total amount of Fe and more than 40% of the total amount of Al were DCB extractable indicating the presence of oxyhydroxides and HA bound forms.

The presence of Na humates and phyllosilicates in the soluble and insoluble fractions, respectively, affected the Na concentration with the increasing of pH value (Table 5).

K was much less abundant than Na. The highest concentrations were found in the insoluble fractions suggesting that besides Na^+, K^+ also neutralized the negative surface charges of the clay minerals. Mg followed more or less the same trend as K. As far as Ca was concerned, no trend could be identified suggesting that this cation is exchangeable on both the phyllosilicate and Ca humates. It can also act as a bridge between the organic and inorganic phases.

DISCUSSION

The HA fraction insoluble at pH 1 might correspond to the HA extracted following the IHSS procedure, except for the purification step. It has the typical composition of a mature HA formed in organic soils in a temperate climate: a high acidic group content and high aromaticity, but a low content of O- and N-containing groups [44]. A compound with nearly the same characteristics is obtained at pH 3. The inorganic phase in the fraction insoluble at pH 1 and pH 3 accounts for about 25% of the total. It is largely constituted by clay minerals (montmorillonite-kaolinite) which only in drastic treatments, including HF-HCl, are able to dissolve. We cannot establish whether these organo-mineral complexes were present in the soil and have been extracted or whether they are formed between two dissolved separated phases. On the other hand, the binding of HA to kaolinite and montmorillonite at pH ≤3 has been assessed by several adsorption studies [45]-[49]. The interaction between humic substances and clay minerals can include specific and columbic interactions, but Schutless et al. [46] recommend being careful in interpreting adsorption studies since at low pH values, adsorption and precipitation mechanisms are difficult to differentiate. The organo-mineral aggregates could also have been stabilized by the presence of Al and Fe, not included in phyllosilicate structures, and low amounts of alkaline and alkaline earth metals. Polyvalent cations are known to promote the interaction of HA with phyllosilicates through the formation of cation bridges [47],[50].

The soluble fractions at pH 1 and pH 3 account for less than 10% of the total material but this data is underestimated because part of this fraction has been lost during the dialysis step. The organic phase contains less than 10% aromatic carbon and is rich in O- and N-containing groups and carboxylic groups. Therefore, it should be composed of soluble polar molecules such as polysaccharides, organic acids, and amino acids. The inorganic fraction is mainly composed of a silica gel-like phase whose formation could be an artifact due to the polymerization of silicic acid by freeze-drying [34]. On the other hand, orthosilicic acid occurs in solutions and on the surface of solid particles in soil [51]. The FT-IR spectra suggest the presence of Al and Fe oxides whose solubility is consistent with the low pH and a possible complexation by small organic molecules.

The fractions insoluble at pH 5 and pH 7 contain a large amount of inorganic material (54% and 61% at pH 5 and 7 respectively). The inorganic phase mainly consists in phyllosilicates (kaolinite and montmorillonite).

Less than 5% Al is DCB extractable indicating that all the Al atoms are included in phyllosilicate structures.

These results suggest that the material precipitated at pH 5 and 7 does not exhibit a typical humic structure. It is more likely an organo-mineral association of clay minerals bound to polar molecules. This result is in agreement with the study by Yong and Mourato [52], performed on the interaction of kaolinite and dextran, which demonstrated the capability of polysaccharides to act as flocculants of clay minerals.

The soluble fractions at pH 5 and pH 7 account for 73% and 90% of the total humic material, respectively. These fractions are particularly rich in aromatic and aliphatic groups and exhibit a very low polar group content as indicated by both the NMR results and the low O/C ratio. The fact that most of the polar molecules were no longer present in this typical humic material, but precipitated as organo-mineral associations (see the above paragraph), suggests that contrary to the usual assumption, they were not strongly bound to the humic acid.

There is evidence of the presence of kaolinite and montmorillonite whose solubility, probably as a colloidal solution, is promoted by the presence of large amounts of Na^+. On the other hand, the phyllosilicates account for only part of the inorganic phase since more than 40% Al and 70% Fe are DCB extractable, indicating the presence of oxihydroxides and/or Al and Fe humates. The increased solubility of the humic fraction at increasing pH is expected due to the deprotonation of the carboxylic and phenolic groups. The carboxyl groups are known to be completely undissociated at pH <3 and mostly ionized at pH >7, while the dissociation of the hydroxyl phenolic groups occurs at higher pH values. It is surprising that this fraction, whose organic phase has the typical features of mature humic acid, rich in aromatic and aliphatic moieties and therefore largely hydrophobic, was soluble at pH 5.

Our results are in contrast with the reported data concerning the solubility of HA which are usually found to be completely soluble at much higher pH (at least >8) [53]. One of the reasons for such a discrepancy could be the solubilization procedure which several studies report as affecting the behavior of the humic substances. Klucakova et al. [54] have observed that the fresh, wet HA precipitate is very easily peptized upon washing with water which is accompanied by a gradual increase in pH. Solubility of dried, solid preparation in water is much more difficult, especially when not freeze-dried. This last observation is consistent with the experiments by Brigante et al. [12] which observed a complete solubilization only at pH ≥ 9 (*eliminato un pezzo perché usato nell'introduzione*). The discrepancy between our results and those of Brigante et al. [12] could be explained by the different procedure for modifying the pH. In our experiment, we started from a humate solution to which amounts of HCl were added, while Brigante et al. [12] started from a solid humic acid brought to higher pH by NaOH addition. This suggests that it is less easy to form aggregates by acidifying a humate solution than to disrupt them by bringing solid HA to higher pH. Therefore, the solubilization-precipitation process of HS should behave hysteretically.

Our study points out that a decrease in the pH of a humate solution from 5 to 3 provokes a drastic change in the aggregation of the humic molecules. This pH range has not yet been fully investigated. The average pK_a of the carboxylic groups of the HA is 4.5 [55],[56]. More recent studies [57], based on the Anderson-Hasselbach interpretation, have established that the apparent pK_a of a peat humic acid was 4.93. Alvarez-Puebla and Garrido [58] measured the zeta potential of a grey humic acid against pH and observed a drop in the pH intervals 3.6 to 4.2 and 5.9 to 6.2, with a section of null slope between the two. The null slope section between these drops seems to indicate that the electrokinetic properties of the colloids do not change in this interval. The plot of the size variation against pH showed great similarity. The porosity measured through CO_2adsorption isotherms indicated a decrease in the pore size and surface area between pH 4 and pH 5, while no relevant difference was observed between pH 2 and pH 4. The results of Alvarez-Puebla and Garrido [58] indicate that the decrease in colloidal size appears to be more influenced by the ionization of phenolic groups than by the ionization of carboxylic groups. This is because the phenolic groups and the undissociated carboxylic groups can participate in forming hydrogen bonds thus promoting the coiling of the molecules and the formation of aggregates.

Another factor affecting the solubility of HS is the amount and composition of the inorganic phase. Extracted HA are usually treated with HCl-HF solution until a low ash content is reached, usually $<2\%$. Little attention is given to the residual inorganic phase because it is assumed that it does not affect the behavior of the organic material. Klucakova et al. [59] reported the equilibrium constants concerning lignitic HA treated or not treated with HCl-HF. From these results, it appears that the ratio between HA(aq) and HA(s) decreased as ash content increased.

To better understand the behavior of the HA at pH 5, especially the possibility of forming metallic humates, it would be useful to have more information on the nature of the carboxylic groups. Powell and Fenton [53] performed a titration of the carboxylic groups of a peat HA. They found that up to 20% of the COOH groups were weak acids, with titration constants above log $K=5.0$, indicative of polyprotic moieties which contribute significantly to the metal binding properties of HS. This suggests that the dissociated COOH groups below pH 5 are monoprotic moieties unable to bind bivalent and trivalent cations. The soluble fractions at pH 5 and 7 contain more than 20% of inorganic material probably present as a colloidal solution whose stability is promoted by the sodium ions. According to the results of

Powell and Fenton [53], the formation of metal humates with bivalent and trivalent ions could take place at pH 7. On the other hand, the complete ionization of the carboxylic groups at this pH could explain the solubility of the corresponding fraction.

CONCLUSIONS

Our results indicate a notable change in the behavior and composition of the humic acid we studied in the pH range from 3 to 5. In this pH range, the humic acid changes from insoluble to a colloidal solution, notwithstanding the hydrogen bonds and weak molecular forces. On the other hand, it has also been demonstrated [54],[59] that solubilization of HA is not only dependent on the deprotonation of COOH and that undissociated molecules can be water soluble.

Most of the fraction consisting in O-containing groups, notwithstanding its high polarity, does not remain in solution at pH 5 and pH 7 but precipitates with the phyllosilicate-rich inorganic phase. This suggests that it is not strongly bound to the humic fraction.

At pH 1 and pH 3, the inorganic phase bound to the insoluble humic material was largely constituted by clay minerals and some Al and Fe hydroxides. Moreover, at lower pH values, the soluble fractions isolated were very poor and composed of a silica gel-like associated with polar organic material.

The results emphasize the effect that the preparation procedure can have on the behavior of the humic substances.

AUTHORS' CONTRIBUTIONS

The work presented here was carried out in collaboration between all authors. MN defined the research theme; MN and AB designed methods and experiments; DV carried out the laboratory experiments, AB, MG and MN analyzed the data, interpreted the results, AB and MN wrote the paper. All authors have contributed to, seen, and approved the final manuscript.

REFERENCES

1. Lal R (2004) Soil carbon sequestration impacts on global climate change and food security. Science 30:1623–1627
2. Piccolo A, Spaccini R, Nieder R, Richter J (2004) Sequestration of a biologically labile organic carbon in soils by humified organic matter. Clim Change 67:329–343
3. Nardi S, Muscolo A, Vaccaro S, Baiano S, Spaccini R, Piccolo A (2007) Relationship between molecular characteristics of soil humic fractions and glycolytic pathway and Krebs cycle in maize seedlings. Soil Biol Biochem 39:3138–3146
4. Canellas LP, Zandonadi DB, Busato JG, Baldotto MA, Simoes ML, Martin-Neto L, Facanha AR, Spaccini R, Piccolo A (2008) Bioactivity and chemical characteristics of humic acids from tropical soils sequence. Soil Sci 173:624–637
5. Nebbioso A, Piccolo A (2009) Molecular rigidity and diffusivity of Al^{3+} and Ca^{2+} humates as revealed by NMR spectroscopy. Environ Sci Technol 43:2417–2424
6. Nègre M, Boursier C, Abbate C, Baglieri A, Gennari M (2008) Use of soil colloids to evaluate adsorption of phenanthrene and its mobilization by different solutions. J Environ Sci Health A 43:443–451
7. Baglieri A, Abbate C, Borzì D, Gennari M (2009) Adsorption of pyrimethanil on soil and some of its colloids. Fresenius Environ Bull 18:578–584
8. Hayes MHB, Swift RS, Wardle RE, Brown JK (1975) Humic materials from an organic soil: a comparison of extractant and of properties of extracts. Geoderma 13:231–245
9. Swift RS (1996) Organic matter characterisation. In: Methods of soil analysis. Part 3. Chemical methods Soil Science Society of America and American Society of Agronomy, Madison pp 1011–1069
10. Piccolo A, Conte P (2000) Molecular size of humic substances. Supramolecular associations versus macromolecular polymers. Adv Environ Res 3:508–521
11. Sutton B, Sposito G (2005) Molecular structure in soil humic substances. Environ Sci Technol 39:9009–9015

12. Brigante M, Zanini G, Avena M (2007) On the dissolution kinetics of humic acid particles: effects of pH, temperature and Ca^{2+} concentration. Colloids Surf A: Physicochem Eng Aspects 294:64–70

13. Choudhri MB, Stevenson FJ (1957) Chemical and physicochemical properties of soil humic colloids: III. Extraction of organic matter from soil. Soil Sci Soc Am Proc 21:508–513

14. Gascho GJ, Stevenson FJ (1968) Improved method for extracting organic matter from soil. Soil Sci Soc Am Proc 32:117–119

15. Nègre M, Vindrola D, Spera S, Ferraris L, Gennari M (2002) Effect of the chemical composition of soil humic acids on their viscosity, surface pressure and morphology. Soil Sci 167:636–651

16. Sanders JKM, Hunter BK (1993) Modern NMR spectroscopy: a guide for chemists Oxford University Press, New York

17. Mehra OP, Jackson ML (1960) Iron oxide removal from soils and clays by a dithionite-citrate system buffered with sodium bicarbonate. Clays and Clay Minerals. 7:317–327 Proceedings of the 7th National Conference on Clays and Clay Minerals, Washington, DC

18. Piccolo A (1988) Characteristics of soil humic extracts obtained by some organic and inorganic solvents and purified by HCl-HF treatment. Soil Sci 146:418–426

19. Perdue EM (1985) Acidic functional groups of humic substances. In: Aiken GR, McKnight DM, Wershaw RL, McCarthy P (ed) Humic substances in soil, sediment, and water: geochemistry, isolation and characterization, Wiley, New York pp 493–526

20. Stevenson FJ (1994) Extraction, fractionation, and general chemical composition of soil organic matter. In: Stevenson FJ (ed) Humus chemistry. Genesis, composition, reactions, Wiley, New York pp 26–54

21. Hatcher PG, Schnitzer M, Dennis LW, Maciel GE (1981) Aromaticity of humic substances in soils. Soil Sci Soc Am J 45:1089–1094

22. Schnitzer M, Preston CM (1986) Analysis of humic acids by solution and solid-state carbon-13 nuclear magnetic resonance. Soil Sci Soc Am J 50:326–331

23. Mao JD, Hu WG, Schmidt-Rohr K, Davies G, Ghabbour EA, Xing B (2000) Quantitative characterization of humic substances by solid-state carbon-13 nuclear magnetic resonance. Soil Sci Soc Am J 64:873–884
24. Yanagi Y, Tamaki H, Otsuka H, Fujitake N (2002) Comparison of decolorization by microrganisms of HA with different C-13 NMR properties. Soil Biol Biochem 34:729–733
25. Spaccini R, Mbagwu JSC, Conte P, Piccolo A (2006) Changes of humic substances characteristics from forested to cultivated soils in Ethiopia. Geoderma 132:9–19
26. Piccolo A, Conte P, Spaccini R, Mbagwu JSC (2005) Influence of land use on the characteristics of humic substances in some tropical soils of Nigeria. Eur J Soil Sci 56:343–352
27. Saab SD, Martin-Neto L (2007) Condensed aromatic rings and E4/E6 ratio: humic acids in gleysoils studied by NMR CP/MAS C-13 and dipolar dephasing. Quim Nova 30:260–263
28. Forte C, Piazzi A, Pizzanelli S, Certini G (2006) CP MAS ^{13}C spectral editing and relative quantitation of a soil sample. Solid State Nucl Mag 30:81–88
29. Mathers NJ, Xu Z, Berners-Price J, Senake Perera MC, Saffigna PG (2002) Hydrofluoric acid pre-treatment for improving ^{13}C CPMAS NMR spectral quality of forest soils in south-east Queensland, Australia. Aust J Soil Res 40:655–674
30. Piccolo A, Campanella L, Petronio BN (1990) Carbon-13 nuclear magnetic resonance spectra of soil humic substances extracted by different mechanisms. Soil Sci Soc Am J 54:750–756
31. Stevenson FJ (1982) Humus chemistry: genesis, composition, reactions Wiley, New York
32. Francioso O, Ciavatta C, Montecchio D, Tugnoli V, Sàncez-Cortés S, Gessa C (2003) Quantitative estimation of peat, brown coal and lignite humic acids using chemical parameters, ^{1}H-NMR. Bioresource Technol 88:189–195
33. van der Marel HW, Beutelspacher H (1976) Atlas of infrared spectroscopy of clay minerals and their admixtures Elsevier, Amsterdam
34. Madejovà J, Komadel P (2001) Baseline studies of the clay minerals society source clay: infrared methods. Clay Clay Mineral 49:410–432

35. Vinkler P, Lakatos B, Meisel J (1976) Infrared spectroscopic investigation of humic substances and their metal complexes. Geoderma 15:231–242
36. Piccolo A, Stevenson FJ (1982) Infrared spectra of Cu^{2+}, Pb^{2+}, and Ca^{2+} complexes of soil humic substances. Geoderma 27:195–208
37. Inoue K, Huang PM (1985) Influence of citric acid on the formation of short-range ordered aluminosilicates. Clay Clay Mineral 33:312–322
38. Inoue K, Huang PM (1990) Perturbation of imogolite formation by humic substances. Soil Sci Soc Am J 54:1490–1497
39. Bartoli F, Philippy R, Portal JM, Gerard B (1992) Poorly ordered Fe oxides, colloidal dispersion and soil aggregation. I. Effect of humic macromolecules on surface and colloidal properties of Fe(III) polycations. J Soil Sci 43:47–58
40. Leone P, Nègre M, Gennari M, Boero V, Celis R, Cornejo J (2001) Adsorption of imidazolinone herbicides on ferrihydrite-humic acid associations. J Environ Health, Part B 2:127–142
41. Cornejo J (1987) Porosity evolution of thermally treated hydrous ferric oxide gel. J Colloid Interf Sci 115:260–263
42. Hsu PH (1989) Aluminum hydroxides and oxyhydroxides. In: Dixon JB, Weed SB (ed) Minerals in soil Environments, Soil Science Society of America, Madison pp 331–378
43. De Cristofaro A, He JZ, Zhou DH, Violante A (2000) Adsorption of phosphate and tartrate on hydroxy-aluminum-oxalate precipitates. Soil Sci Soc Am J 64:1347–1355
44. Watanabe A, Takada H (2006) Structural stability and natural ^{13}C abundance of humic acids in buried volcanic ash soil. Soil Sci Plant Nutr 52:145–152
45. Tombacz E, Gilde M, Abraham I, Sza'nto' F (1988) Effect of electrolyte concentration on the interaction of humic acid and humate with montmorillonite. Appl Clay Sci 3:31–52
46. Schulthess CP, Huang CP (1991) Humic and fulvic acid adsorption by silicone and aluminium oxide surfaces on clay minerals. Soil Sci Soc Am J 55:34–42
47. Elfarissi F, Pefferkorn M (2000) Kaolinite: humic acid interaction in the presence of aluminium ion. Colloids Surf A 168:1–12

48. Tombacz E, Libor Z, Illes E, Majzik A, Klump E (2004) The role of reactive surface sites and complexation by humic acids in the interaction of clay mineral and iron oxide particles. Org Geochem 35:257–267
49. Nègre M, Leone P, Trichet J, Defarge C, Boero V, Gennari M (2004) Characterization of model soil colloids by cryo-scanning electron microscopy. Geoderma 121:1–16
50. Feng X, Simpson AJ, Simpson M (2005) Chemical and mineralogical controls on humic acid sorption to clay mineral surfaces. Org Geochem 36:1563–1566
51. Wada SI, Wada K (1980) Formation, composition and structure of hydroxy-aluminosilicate ions. J Soil Sci 31:457–467
52. Yong R, Mourato ND N (1990) Influence of polysaccharides on kaolinite structure and properties in a kaolinite-water system. Can Geotechnical J 27:774–788
53. Powell HJK, Fenton E (1996) Size fractionation of humic substances: effect on metal binding properties. Anal Chim Acta 334:27–38
54. Klucakova M, Peka M (2008) Behaviour of partially soluble humic acids in aqueous suspensions. Colloids Surf A 318:106–110
55. Christensen JJ, Hansen LD, Izatt RM (1976) Handbook of proton ionization heats and related thermodynamic quantities Wiley, New York
56. Martell AE, Smith RM (1977) Critical stability constants: Other organic ligands, vol 3 Plenum, New York
57. Terashima M, Fukushima M, Tanaka S (2004) Influence of pH on the surface activity of humic acid: micelle-like aggregate formation and interfacial adsorption. Colloids Surf A 247:77–83
58. Alvarez-Puebla RA, Garrido JJ (2005) Effect of pH on the aggregation of a gray humica acid in colloidal and solid states. Chemosphere 59:659–667
59. Klucakova M, Peka M (2005) Solubility and dissociation of lignitic humic acids in water suspension. Colloids Surf A 252:157–164

Chapter 5

Geochemical Modeling and Multivariate Statistical Evaluation of Trace Elements in Arsenic Contaminated Groundwater Systems of Viterbo Area, (Central Italy)

Giuseppe Sappa, Sibel Ergul, and Flavia Ferranti
Michèle Negre[2]

Dipartimento di Ingegneria Civile, Edile ed Ambientale, Sapienza, Università di Roma, Via Eudossiana, 18, 00186 Rome, Italy

ABSTRACT

Contamination of groundwater by naturally occurring arsenic has recently become a disturbing environmental problem in Viterbo area, Central Italy. Arsenic concentrations in most of the public supply networks exceed the maximum allowable limit of 10 μg/l (WHO) for drinking water. The primary purpose of this paper is to obtain a better understanding of the factors contributing to the high levels of As in water supply networks. This study focuses on (a) the determination of basic hydrochemical characteristics of groundwater, (b) the identification of the major sources and processes controlling the As contamination in public supply networks, (c) to find out possible relationships among the As and other trace elements through principal component analysis (PCA). Groundwater samples from public water supply wells and springs were collected and analysed for physico-chemical parameters and trace elements. Springs and well water samples are predominantly of the Na–HCO_3, Na –Ca–HCO_3 and Ca–HCO_3 types and the highest arsenic concentrations were observed in Na–HCO_3 type water. Eh-pH diagrams reveal that $H_2AsO_4^-$ and $HAsO_4^{2-}$, As (V) arsenate, are the dominating As species highlighting slightly to moderately oxidizing conditions. Geochemical modeling indicates that arsenic-bearing phases were undersaturated in the groundwater, however most of the samples were saturated with respect to Fe (i.e. magnetite, hematite and goethite) and Al (diaspore and boehmite) oxide and hydroxide minerals. Concentrations of As, Li, B, Co, Sr, Mo, U and Se are highly correlated ($r > 0.7$) with each other, however in some groundwater samples As show also good correlations ($r > 0.5$) with Fe and Mn elements reflecting the relationships among the trace elements result from different geochemical processes. Evaluation of the principal component (PCA) analysis and geochemical modeling suggest that the occurrence of As and other trace element concentrations in groundwater are probably derived from (i) weathering and/or dissolution of volcanic source aquifer materials and (ii) adsorption/desorption processes on the Fe and Al oxide and hydroxide minerals.

INTRODUCTION

Groundwater resources are generally less susceptible to pollution than surface water and are considered the best supply for drinking water, however in many parts of the world people suffer from poor drinking water quality (Chapman 1992; Foster et al. 1997). The quality of groundwater depends on the composition of recharging water and the mineralogy of the geological formations in the aquifers. The impact of human activities and the environmental parameters may also affect the geochemical mobility of certain constituents in groundwater (Chenini and Khmiri 2009). In Italy, 85% of drinking water is drawn from underground sources, especially in Viterbo area, central Italy, 94% of water used in this area rely on groundwater in the aquifers. In the last decades, the attention of legislators have increased on water and groundwater sources quality, in particular, many laws have been enacted with the purpose of preventing or at least mitigate the presence of pollutants in water. Recently, in the province of Viterbo, Central Italy, the government declared a "state of emergency" due to the presence of "high levels" of arsenic in drinking water. Preliminary initiatives have been undertaken by Italian National Institute of Health (ISS) and Optimal Territorial Area Authorities n°1 (ATO) to develop technical solutions, plant and/or management to reduce the arsenic concentrations in groundwater used for domestic purposes. The presence of arsenic and other toxic trace elements in water supply networks is a threat to population and agricultural activities (Roychowdhury et al. 2002; Meharg and Rahman 2003Shakeel and Amal 2011). The annual reports of National Research Council (1999 2001) affirmed the limit of arsenic in drinking water at 50 μg/l, however, the US federal drinking water standard, or maximum contaminant level (MCL), was brought down to 10 μg/l. According to the World Health Organization (WHO), the provisional limit of arsenic in drinking water is 10 μg/l, and the same limit was adopted by the European commission (WHO 2006). The same limit has been take place in the Italian legislation (Legislative Decree 31/2001 "Implementation of Directive 98/83/EC on the quality of water intended for human consumption").

Investigations in the last few years have shown that As mobilization can occur in many aquifers and concentrations can exceed the drinking-water quality standards in different hydrogeological conditions. Mechanisms responsible for these high As concentrations have been reported in many studies. High concentrations of naturally occurring As, in groundwater, is associated with the presence of geothermal systems and/or volcanic-sedimentary rock aquifers (Ballantyne and Moore1988; Webster and Nordstrom 2003), and the mobilization of As is still open to many interpretations (Casentini and Pettine 2010). Volcanic degassing, interaction with deep-rising fluids and leaching of ore deposits may also important factors influencing the natural arsenic enrichment in groundwater circulating in active volcanic areas and geothermal fluids (Piscopo et al. 2006; López et al. 2012). Arsenic occurrence can take place through a combination of natural processes (e.g., weathering reactions, biological activity, leaching process, redox conditions in the subsurface environment, different water–rock interactions) as well as anthropogenic activities including coal mining and its combustion (Charlet and Polya 2006; Smedley and Kinniburgh 2002; Bose and Sharma 2002; Nickson et al. 2000). Arsenic is predominantly released from rocks with primary or secondary As or As-bearing minerals due to physical, chemical or microbiological weathering into aqueous environments. To understand (As) enrichment in water systems, the identification of geochemical parameters are the most important tools to better understand the occurrence and genesis of high As concentrations in groundwater, which are predominantly due to release from geogenic resources. The study of major and trace elements also allows us to evaluate water quality indices which are the important parameters for public health (Anawar et al. 2003Barbieri et al. 2013). This will help to identify the origin and the processes leading the high concentrations of toxic elements, and hence to develop possible mechanisms of their removal from solution, providing at the same time new approaches on the quality of groundwater (Bratus et al. 2006; Parisi et al. 2011; Vivona et al. 2006; Kumar et al. 2010).

This paper presents an integrated study on the occurrence and the distribution of arsenic and other trace elements in public water supply networks of Viterbo area, Central Italy. The main objective of the paper is to understand the control of geochemical processes on the As enrichment in groundwater and its relationship with other trace element concentrations. In order to better understand the implications of the mentioned geochemical processes for water quality, a sampling survey was carried out on drinking water supply networks, which covers 231 different sources including 153 wells and 78 springs. A detailed investigation was carried out on the collected samples based on physico-chemical parameters (pH, temperature, electrical conductivity, etc.), major ion and trace element chemistry. Then, conventional graphical plots, principal component analysis and geochemical modelling techniques were applied to evaluate the geochemistry of As and other trace elements and the mechanisms of their release into groundwater. Results from this work will help to design regional-scale studies of ground-water quality and to find out appropriate remediation techniques minimizing elevated levels of naturally occurring contaminants.

Geological and Hydrogeological Setting of the Study Area

The Province of Viterbo, Central Italy, is located between the Tyrrhenian Sea coast and the Central Apennines mountains, shown in Figure 1., and was formed by two different volcanic activities in the late Pliocene - Pleistocene period: the acidic volcanic cycle of Tuscany Magmatic Province and K-alkaline volcanic cycle of the Roman Magmatic Province (Barberi et al. 1994). These volcanic formations host several aquifer systems due to the high porosity and permeability that characterize volcanic rocks. In the study area, continuous and generally unconfined volcanic and discontinuous several perched aquifers have been found. Most of the springs are related to the perched aquifers and generally discharge less than 0.01 m^3/s. The Pliocene–Quaternary magmatic activity in the peri-Tyrrhenian sector of Italy has produced several geothermal anomalies (Della Vedova et al. 1984; Mongelli et al. 1989). The western side of Viterbo town reserves several thermal springs, which were known in Roman times and, nowadays, they are still exploited. The thermal springs of Viterbo area are related to the regional circulation of groundwater in the Mesozoic limestone aquifer (Minissale and Duchi 1988). The thickness of the volcanic aquifer (i.e. fresh waters) decreases in the thermal area where it includes more layers of travertine deposits (Piscopo et al. 2006). However, our paper considers the hydrogeochemical data of springs and shallow aquifers (temperature ranges from 10 to 25°C) currently supplying drinking water networks of Viterbo area. The presence of this important thermal groundwater circuit influences the hydrochemical properties of fresh groundwater, used for drinking supply. Three different geological districts emerge within the study area from north to south: the Vulsino district, around Bolsena Lake, located in the northernmost part, the Cimino-Vico district around Vico Lake and the Sabatino district near to lake of Bracciano (Figure 1). The outcropping rocks include mainly volcanic and volcanoclastic formations, different in age and chemical composition.

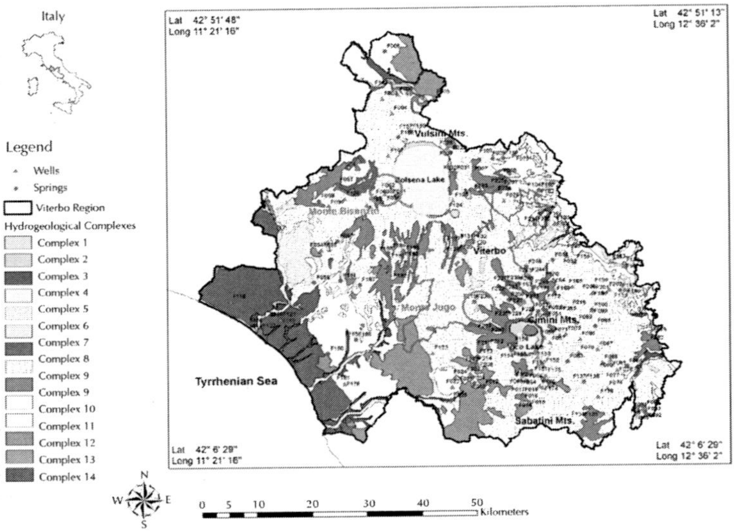

Figure 1: Simplified hydrogeological map of study area and location of springs and wells: (1) Recent deposits (Oligocene); (2) Detritic complex (Pleistocene-Oligocene); (3) Alluvial complex (Pleistocene-Oligocene); (4) Alluvial deposits (Pleistocene-Oligocene); (5) Travertines (Pleistocene-Oligocene); (6) Sand dunes (Pleistocene-Oligocene); (7) Fluvial lacustrine deposits (Oligocene); (8) Pyroclastic complex (Pliocene-Pleistocene); (9) Lavas and lithoidal ignimbrites; (Pliocene-Pleistocene); (10) Heterogeneous clastic deposits (Pleistocene); (11) Pliocene Clays; (12) Clayey-marly Flysch complex with interbedded lithoids (Cretaceous-Miocene); (13) Cretaceous pelagic limestone (Cretaceous); (14) Metamorphic complexes (Palaeozoic).

The Vulsini volcanic district is located at the northwestern end of the potassic Roman co-magmatic region, which was developed along the Tyrrhenian coast of Central Italy during Quaternary (Figure 1). The Vulsini volcanic complex include entire series of potassic rock types, with a predominance of trachytes and phonolites in terms of erupted volumes and mostly characterized by Plinian pumice-fall and ash-pumice flow deposits from larger explosive eruptions (Palladino et al. 2010). The volcanic complexes show high permeability due to the high porosity and fissures. These permeable terrains infiltrate meteoric waters and feed aquifers, located at different depths. Vulsino Basin was formed by the following hydrogeological units: lacustrine and fluvial alluvial deposits, pyroclastics, lavas, lithoid ignimbrites and volcanic-sedimentary deposits. The main shallower aquifer is located in the volcanic deposits and rests on a clay substratum consisting relatively fresh and low salinity waters. The lithotypes with low permeability, composed of sedimentary formations characterized by shallow marine to continental clays, sands, and conglomerates (Upper Miocene – Quaternary) and Upper Cretaceous – Oligocene flysch formations, which correspond to an aquiclude separating at least two main reservoirs. The Vulsino volcanic aquifer feeds the Bolsena Lake, which is the biggest volcanic lake in Europe. The deeper aquifer, confined between the impermeable formations (Ligurides and Neogene marine sediments) underlying the volcanic products and the less permeable volcanic rocks, is characterized by relatively high salinity thermal waters (Pagano et al. 2000).

The Cimino and Vico volcanic districts are different from each other, both their evolution and type of the magmas they produced. The volcanism in Cimino district is related to the acidic-felsic cycle of the Tuscany Magmatic Province consisting SiO_2 rich magma, while Vico district shows the K-alkaline cycle volcanism of the Roman Magmatic Province (Coli et al. 1991; Perini et al. 2000). The Vico complex, located at south part of the Cimino volcanic complex, consists of a strato-volcano with a central caldera depression housing lake Vico (Figure 1). This complex was characterized by explosive volcanic activity, which was developed between 0.8 and 0.4 Ma ago. The products of Vico volcanic activity include leucitites, phono-tephrites and leucite-phonolites, while Cimino complex are mainly composed of latites and trachytes (Borghetti et al. 1983). The basement of Cimini and Vico volcanic constituted by sedimentary rocks, the Upper Cretaceous-Oligocene flysch and the Triassic- Paleogene carbonate rocks, (Cimarelli and De Rita 2006). According to previous studies, a continuous volcanic aquifer, discharges mainly into streams and springs, and several limited discontinuous perched aquifers are found in the area (Boni et al. 1986; Capelli et al. 2005; Baiocchi et al. 2007). The mean yield of the volcanic aquifer has been estimated to be between 5 and 7 m3/s. The aquifer system of Cimini and Vico volcanic area is limited by the Pliocene-Pleistocene sedimentary complex on its eastern edge and by the Upper Cretaceous-Oligocene Flysch on its western and south-western sides (Figure 1). A second deeper carbonate aquifer, located in the thick sequence of Mesozoic limestones, has been also found in the area, which consist of the Triassic-Palaeogene carbonate rocks hosting a thermal reserviour (Chiocchini et al. 2010Baiocchi et al. 2012). The volcanic aquifer rests on the carbonate aquifer and separated by low-permeability Pliocene-Pleistocene and Upper Cretaceous-Oligocene sedimentary rocks (Baiocchi et al. 2007). The thermal water of carbonate aquifer is characterized by high salinity and temperature. In the west part of Viterbo, the volcanic basement has been uplifted and reduced the thickness of sedimentary rocks. The thermal waters rise up from the carbonate aquifer through the normal faults and joints and undergo mixing with cold waters (Piscopo et al.2006).

The Sabatini volcanic district belongs to the potassic Roman comagmatic region of quaternary age that extends along the Tyrrhenian coast (Central Italy) (Figure 1). The location of the major volcanic districts along a NW–SE tectonic trend is linked to the extensional faulting started during the Pliocene that is related to the opening of the Tyrrhenian back arc basin (De Rita et al. 1983). This area characterized by the presence of numerous cold and thermal waters and CO_2- rich gas emissions due to the post-orogenic magmatic activity that occurred from Pliocene to Quaternary, in response to tectonic movements associated with the opening of the Tyrrhenian Sea (Minissale 2004). The main hydrogeological patterns are related to different aquifers: 1) a deeper one located in Mesozoic anhydritic-carbonate formations, and 2) shallow aquifer(s) hosted in the volcanic and sedimentary Plio- Quaternary deposits which locally may show a relatively high permeability (Dall'Aglio et al.1994).

METHODOLOGY

In the present research, the existing monitoring data, obtained from Italian National Institute of Health (ISS), for 231 individual domestic water supply wells and springs, was employed for geochemical modeling and statistical analysis to identify the occurrence and distribution of arsenic in the aquifers of Viterbo area. Hydrogeochemical characterization of groundwater was evaluated by means of physico-chemical analysis on the collected samples to identify the chemical characteristics and their relation with existing quality of groundwater of each municipality. The study approach includes conventional graphical plots and principal component analysis (PCA) of the hydrochemical data to define the geochemical evaluation of groundwater based on the ionic constituents, hydrochemical facies with distinct characteristics and factors controlling groundwater quality. During sampling, from 2007 to 2009, physico-chemical parameters of 231 groundwater samples (i.e. T, EC, TDS and pH) were determined in the field using PC 300 Waterproof Hand-held meter. These samples were analyzed only for major ions, Arsenic and Fluoride concentrations. The analysis were carried out at the Geochemistry Laboratory of Sapienza University of Rome and in the laboratory of National Institute of Health (ISS). Water samples were filtered through cellulose filters (0.45 μm). Each sample was divided into two subsamples: the first had stored at 4°C and been used to determine their major and minor constituents, with a Dionex DX-120 ion chromatograph (reliability ±2%). A Dionex CS-12 column was used for determining cations (Na^+, K^+, Mg^{2+}, Ca^{2+}), while a Dionex AS9-SC column was used for anions (SO_4^{--} HCO_3^-, Cl^-, NO_3^-). The analytical accuracy of these methods ranged from 2% to 5%. Bicarbonate content was measured by titration with 0.1 N HCl using colour turning method with methyl orange as indicator.

To understand As enrichment and its relationship with other trace elements, a new sampling survey was carried out in 2012. However, trace element concentrations were measured from only the seven most biggest and important drinking water supply wells in the study area. These wells are located in Monte Bisenzio area, near Vulsino district, and Monte Jugo area belonging to Cimino Vico districts (Figure 1). The concentration of trace elements were measured at different pumping rates (27) on the groundwater samples from Monte Jugo and Monte Bisenzio areas. Groundwater samples were inserted acid-washed polyetheylene bottles and acidified with concentrated nitric acid (Ultrapur, Merck, v/v) to pH<2 and stored at 4°C and analyzed by inductively coupled plasma mass spectrometry, ICP-MS, Plasmaquad 3 Vg Elemental, (reliability ±2%) to identify trace elements. The relationship between the concentrations of arsenic were correlated with major ions and trace elements measured in groundwater to identify the source and mechanism of arsenic release in the aquifers systems of Viterbo area. All of the available arsenic data sets have been collected and incorporated into a Geographical Information System (GIS) system for the production of environmental contamination map to highlight distribution of As in monitored wells and springs (Figure 2). Garmin eTrex 20 GPS device was used for field data collection, which generally has an accuracy of ±4 m. For the identification of hydro-chemical facies Geochemistry Software AqQA) was employed. Trilinear diagram of

Piper (1944) diagrams has been used to define different hydrochemical facies (Piper 1944). The chemical analysis data of the spring and well water samples have been plotted on the Piper diagram. The PHREEQC software was provided, using WATE4QF database, to compute aqueous speciation and fluid-mineral equilibrium. The estimated saturation indexes of relevant minerals (SI) are approximate due to analytical and activity concentration uncertainties; they are assumed to be ± 0.5 accurate (Parkhurst and Appello 1999). Different chemical parameters, including pH, Eh, temperature, electrical conductivity (EC), total dissolve solids (TDS), Ca, Mg, Na, K, HCO_3, Cl, SO_4, NO_3, F, As and other trace elements (such as Li, V, Cr, Fe, Se, Mo, Zn, Co, Mn, Al), were used in the calculations. Principal component analysis (PCA) was applied to reduce the data sets and to identify relationships among the variables responsible for the source of groundwater contamination.

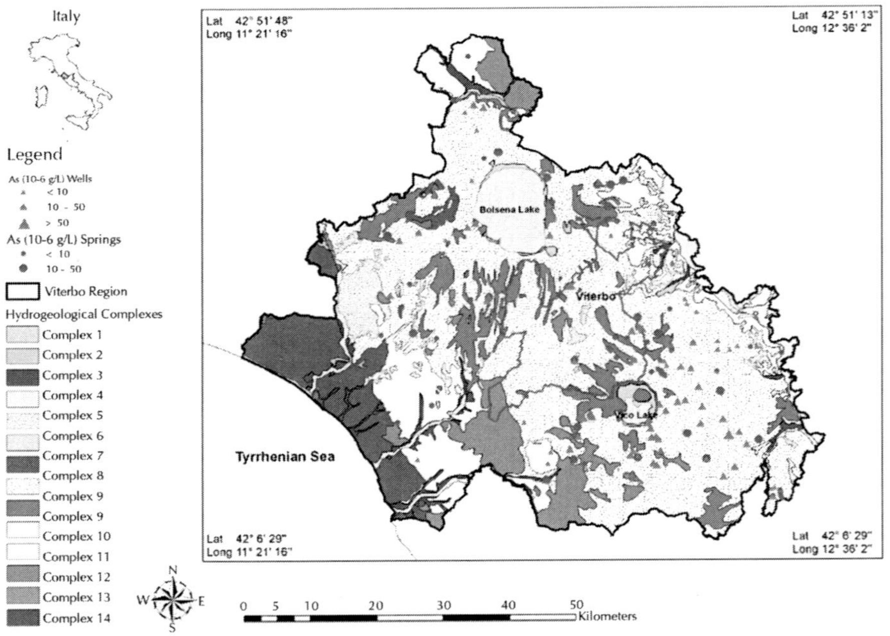

Figure 2: Arsenic distribution in springs and groundwater.

RESULTS AND DISCUSSION

Physicochemical Parameters and Major Constituents

Summary statistics of physicochemical parameters (mean, median, maximum, minimum values and standard deviation values) of well and spring water samples and and guideline values of (WHO 2006) for drinking water are shown in Table 1. The pH values of spring and groundwater samples range from 6 to 8.7 indicating slightly acidic to alkaline nature and the values are in the range of WHO guideline limits (6.5 – 9.2). The mean temperature of groundwater and springs range from 15.4 to 17.1. Total dissolved solids (TDS) and electrical conductivity (EC) show a wide variation from 100.2 to 1477.5 mg/l and 134 to 1603 µS/cm in groundwater samples, and from 86 to 715.4 mg/l and 65 to 894 µS/cm in spring samples. Most of the samples show TDS values below 500 mg/l and can be considered as fresh waters, however few samples are classified as brackish water according to the WHO guidelines. The ionic dominance pattern of the water samples for cations and anions is $Ca^{++} > Na^{+} > K^{+} > Mg^{++}$ and $HCO_3^- > Cl^- > SO_4^- > NO_3^-$. However, high NO_3^- concentrations were observed in spring water samples (up to 100 mg/l) exceeding the permissible limit of 50 mg/l of WHO (2006) guideline values for drinking water. Calcium concentrations in groundwater samples vary between 10 and 221 mg/l (mean 42 mg/l Table 1), while in spring water samples range from 7 to 132 mg/l. Magnesium concentrations in both water samples are generally low (range from 3 to–40 mg/l; mean: 10.2 mg/l in groundwater and 7.4 mg/l in springs). The maximum acceptable limits for Mg is 50 mg/l and for the Ca 75 mg/l. The Na and K concentrations vary from 8 to 104 mg/l and 1 to 71 mg/l in groundwater samples and 8 to 55 mg/l and 1-39 mg/l in springs, respectively. All water samples fall within the guideline levels (<200 mg/l) for drinking water. Bicarbonate concentrations exhibit a wide range: in the groundwater samples ranging from 47 to 992 mg/l (mean: 206 mg/l; Table 1), whereas the concentrations are lower in spring water samples (mean: 150 mg/l). The Cl concentrations in the investigated water samples are found in the range of 8-148 mg/l and the highest values were observed in groundwater samples. There are no samples in excess of the permissible limit of 250 mg/l for chloride. The SO_4^- concentration in water samples range from 2 to 137 mg/l with minimum and maximum values, respectively, and all sampled waters are not exceed the permissible WHO guideline value of 250 mg/l. Major ion composition is probably controlled by water-rock reaction (i.e. dissolution of rock forming minerals), where the volcanic formations are the most dominant formations in the area investigated.

Table 1: Summary statistics of physicochemical parameters (mean, median, maximum minimum and standard deviation (SD) values) of spring and well water samples

WHO guideline values	Physico-chemical parameters	Wells					Springs				
		Min	Max	Mean	Median	SD	Min	Max	Mean	Median	SD
75-200	Ca (mg/l)	10.0	221.0	42.0	27.0	38.8	7.0	132.0	32.8	21.5	31.2
50-150	Mg (mg/l)	3.0	40.0	10.2	8.0	6.6	3.0	18.0	7.4	7.0	3.2
200	Na (mg/)	8.0	104.0	22.9	20.0	13.4	8.0	55.0	18.8	17.0	8.3
200	K (mg/l)	1.0	71.0	19.6	20.0	12.3	1.0	39.0	15.3	17.5	9.2
N.S	HCO3– (mg/l)	47.0	992.0	206.4	163.0	138.3	46.0	416.0	150.0	122.5	90.2
250	SO42– (mg/l)	2.0	137.0	15.2	9.0	18.8	3.0	53.0	11.6	8.5	9.6
250	Cl (mg/l)	8.0	148.0	19.9	15.0	17.8	8.0	54.0	17.3	14.0	9.1
1.5	F– (mg/l)	0.0	3.2	0.8	0.8	0.6	0.0	2.7	0.6	0.4	0.6
50	NO3 (mg/l)	1.0	87.0	14.6	13.0	12.2	1.0	100.0	18.4	15.0	18.6

1000	TDS (mg/l)	100.2	1477.5	351.0	287.4	220	86.0	715.4	271.9	224.3	155.7
1500	EC (µS/cm)	134	1603	434.7	372	258.8	65.0	894.0	342.9	293.0	189.2
6.5-9.2	pH	6.0	8.7	7.0	7.0	0.5	6.2	8.1	6.9	6.9	0.4
N.S	T (°C)	10.4	25.5	17.1	16.7	2.3	9.9	21.4	15.4	15.4	2.1
10 µg/l	As (µg/)	1.0	57.0	14.3	11.0	12.6	1.0	45.0	9.0	7.5	9.1
N.S	V (µg/)	1.0	76.0	15.5	12.0	12.3	2.0	51.0	13.1	10.0	10.2
-	Eh (V)	0.23	0.66	0.3	0.28	0.08	0.23	0.58	0.31	0.31	0.06
-	pe	3.9	10.1	5.4	5.4	0.95	4.0	11.4	5.3	4.9	1.3

Sappa *et al.*

Sappa *et al. SpringerPlus* 2014 3:237, doi: 10.1186/2193-1801-3-237

Trace Elements

Trace element studies were carried out by many researchers on contaminated waters in order to determine the origin of the pollution (Drever 1997; Langmiur 1996; Appelo and Postma 1993). Understanding the distribution of arsenic and other related toxic trace elements in drinking water is essential to identify contamination mechanism and to the develop suitable remediation technologies identifying high-risk areas (Parisi et al. 2011; Giammanco et al. 1998; Aiuppa et al. 2005; Buschmann et al. 2007). To evaluate the safety of water for drinking, concentrations of various trace elements, including Li, Al, B, Be, Co, Cr, F, As, Cu, Fe, Mn, Mo, Ni, Rb, U, Se, Sr, V, and Zn were determined in the most seven important water supply wells and the results were compared with the guidelines for drinking-water quality established by WHO (2006) (Table 2). A series of various trace elements, including Li, Al, B, Be, Co, Cr, F, As, Cu, Fe, Mn, Mo, Ni, Rb, U, Se, Sr, V, and Zn were determined in the studied groundwater. The concentrations of the elements are summarized in Table 2. The arsenic concentrations cover a wide range and; of the 231 water samples collected most water samples exceed the WHO (2006) guideline value of 10 μg/l for drinking water. The distribution of As concentrations in the studied water samples is shown in Figure 2. In groundwater samples, As concentrations range between 1 and 57 μg/l (mean: 14.3 μg/l), in springs between 1 and 45 μg/l (mean: 9 μg/l) (Table 1). In addition, among the determined analytes, concentrations exceeding the World Health Organization recommended drinking water limits were found for F (>1.5 μg/l) and Fe (>300 μg/l). Nickel values varies between 0.01 and 5.6 μg/l, which is below the drinking water guideline value (20 μg/l). The maximum concentrations of Cu and Zn were 91.5 and 71.5 μg/l, respectively. These values are also below the maximum permissible limit of WHO (2006) guideline values. Cobalt concentrations were also relatively low ranging from 0.28 to 1.82 μg/l. From the groundwater samples investigated, two samples showed elevated value of boron (>800 μg/l), and these values exceed the WHO (2006) drinking water guideline values of 500 μg/l. Vanadium concentration in groundwater ranges from 27.9 to 44.96 mg/l and the highest values were observed in Monte Bisenzio groundwater samples. Uranium concentrations varied between 2.7 and 13.8 μg/l not exceeding the WHO (2006) guideline permissible limits. The Fe concentration varies between 2.4 and 814.5.

In general, Fe concentration was low in the studied groundwater samples and the highest concentration was observed in groundwater sample from Monte Bisenzio well exceeding the WHO (2006) guideline value (300 μg/l). Groundwater samples show also high values of Li (9-66.8 μg/l), Sr (199.5-1228.6 μg/l), Rb (30.9-169.6 μg/l) and Mn (0.16-110.5 μg/l), while the rest of trace element concentrations are generally low including Be (0.35-1.19 μg/l), Al (4.62-18.04 μg/l), Mo (0.7-2.65 μg/l), Cr (0.07-1.39 μg/l) and Se (2.7-4.60 μg/l) (Table 2). The presence of As and other trace elements such as V, Mo, U, B, F, Sr are probably related to the circulation of groundwater in the volcanic formations, which is found in most of the study area. Volcanic rocks, especially ashes, are often implicated in the generation of high-As waters. The occurrence of these elements are presumed to have the same origin derived from volcanic source aquifer materials (i.e. volcanic glass, altered tuffs and sediments). Elevated arsenic concentrations in drinking water supplies in several locations within the United States, Argentina, Greece, Turkey, Chileand Italy have been associated with volcanic rocks and ash-flow tuffs (Casentini et al. 2010; Johannesson and Tang 2009; Welch et al.2000). The volcanic materials, especially tuffs that have undergone different types of alteration including mainly secondary silica, iron/manganese oxides, aluminum hydroxides, and various clay mineral phases. The new mineral phase plays an important role for the mobilization of As and other trace elements in terms of different mechanisms such as dissolution of volcanic glass and adsorption/ desorption processes on the secondary mineral phases. Figures 3a and 3b shows the stratigraphic units of the studied wells in Monte Jugo and Monte Bisenzio areas, respectively. The stratigraphy of the wells are peresnted by well depths. As can be seen from the mentioned figures, the wells mainly composed of tuffs and pyroclastic materials. We conclude that the groundwater flowing through an aquifer composed dominantly of volcanic rocks and the weathered products generated from them are characterized by elevated concentrations of these elements.

Table 2: Statistical summary (mean values) of selected trace elements of groundwater

Monitored wells								
Trace elements (μg/l)	MJ1	MJ2	MB1	MB2	MB3	MB4	MB5	WHO limits (μg/l)
Li	9.8	9.0	66.9	16.2	30.4	66.1	33.3	N.S
Be	0.35	0.45	1.19	0.35	0.44	1.12	0.77	N.S
B	93.0	79.6	832.0	145.1	319.9	840.7	381.4	500
Al	5.4	11.9	18.0	4.6	6.4	7.0	10.3	200
V	33.8	38.5	32.5	40.0	34.3	27.9	44.8	N.S
Cr	0.61	1.4	0.07	0.37	0.20	0.10	0.07	50
Mn	0.16	0.35	110.50	30.71	11.49	99.86	99.11	400
Fe	2.4	11.3	114.1	321.8	4.6	36.8	814.5	300
Co	0.29	0.28	1.84	0.38	0.93	1.82	0.81	N.S
Ni	0.00	5.96	0.92	0.01	0.00	1.91	0.06	20
Cu	0.43	91.55	2.47	0.00	0.00	0.46	0.74	2000
Zn	0.01	16.64	2.76	13.05	0.00	1.03	71.50	3000
As	16.73	14.35	13.09	14.50	7.70	14.11	15.12	10
Se	4.60	3.32	3.49	4.00	2.90	4.19	2.70	10
Rb	30.9	34. 6	169.6	73.7	124.2	165.6	101.3	N.S
Sr	324.9	273.4	1228.77	199.5	586.4	1082.7	532.9	N.S
Mo	2.34	2.17	1.35	1.30	0.70	1.10	2.65	70
U	6.26	2.71	12.87	3.38	7.63	10.65	6.76	15

Sappa *et al.*

Sappa *et al. SpringerPlus* 2014 **3**:237, doi: 10.1186/2193-1801-3-237

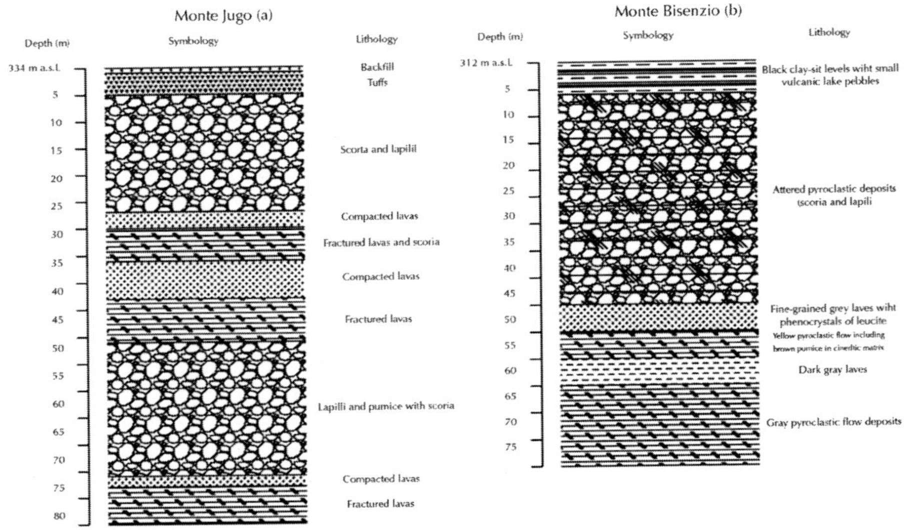

Figure 3: a and b Stratigraphic units of the studied wells.

Hydrochemical Facies & Distribution of Arsenic

The classification of groundwater was studied by plotting the concentrations of major cations and anions in the Piper trilinear diagram to identify hydrogeochemical processes controlling groundwater chemistry (Figure 4). Major ion concentrations in meq/l for each spring and well water samples are reported as percentages of the total anion and cation content (Piper 1944). The trilinear plots suggest that among anions HCO_3^- has a clear dominance. Among cationic species Ca and Na dominate in the groundwater samples. In Figure 4, the plot shows that most of the samples fall within the central part of the cation triangle reflecting the absence of any dominance among the alkali (Na + K) and alkaline earth (Ca + Mg) cations, which indicates the mixing of Ca–HCO_3 and Na-HCO_3 facies. According to Piper diagram, the distribution of hydrochemical facies of groundwater that occur in the Viterbo area can be classified into three groups: a) Facies A-Na- HCO_3 type water, b) Facies B- No dominant type, Na-Ca-HCO_3 (water type in which none of the ions is dominant and/or results of mixing of two or more different facies) and Facies C- Ca –

HCO_3 type. The occurrence of these water types in the aquifers may be due to the interactions between groundwater and different rocks with mineralogical compositions along the groundwater flow paths. From the plots, it is clearly seen that some samples appear in the cation triangle show a tendency towards the alkaline earth composition, however, the water samples which are closer to the Na + K vertex are related to the volcanic aquifers highlighting enrichment in K derived from the alkaline-potassic rocks. The higher concentration of Ca^{2+} Na^{+} and HCO^{3-} in the groundwater may be due to the dissolution of plagioclase feldspars in the rocks resulting in the release of these elements responsible for the various hydro-chemical facies. From the whole samples, 40% of the groundwater from Viterbo province show a composition of Na-HCO_3 type. Besides, 20% of the samples shows Na-Ca-HCO_3 hydro facies, while 40% samples have a composition of Ca-HCO_3. Statistic parameters describing the arsenic distribution for each group of hydrochemical facies of water samples are shown in Figure 5. Evaluation of mean, median, maximum and minimum levels of arsenic depicted in the box plots, indicate the following relationships between hydrochemical facies and arsenic concentrations: (i) Na-HCO_3 water type (Facies A), from wells and springs, are characterized by elevated As concentrations (mean value: $As > 10$ μg/l), exceeding the permissible limit for drinking waters; (ii) The Na-Ca-HCO_3 water type, labelled as Facies B, from wells are characterized by elevated arsenic concentrations above the limit, while the samples from springs show low arsenic contents (mean value: $As < 10$ μg/l); (iii) Ca-HCO_3 water type, Facies C, highlights low arsenic concentrations below the permissible limit (mean value: $As < 10$ μg/l).

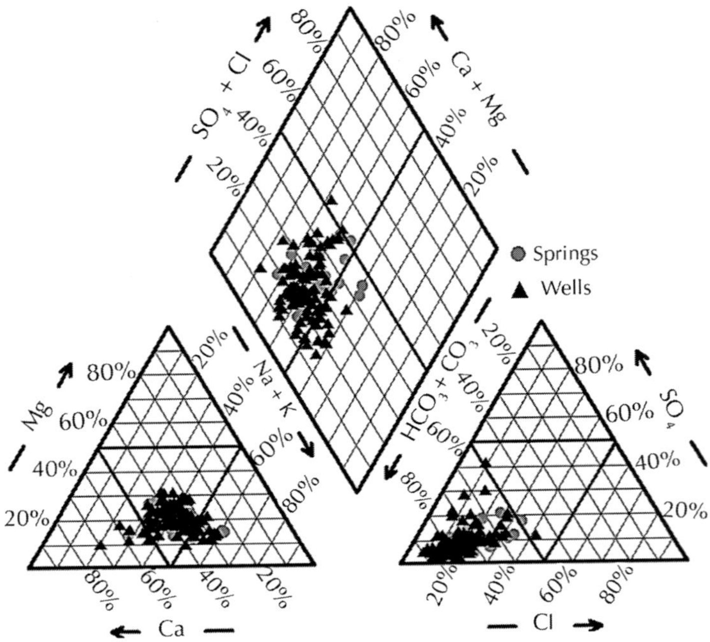

Figure 4: Piper diagram of sampled waters.

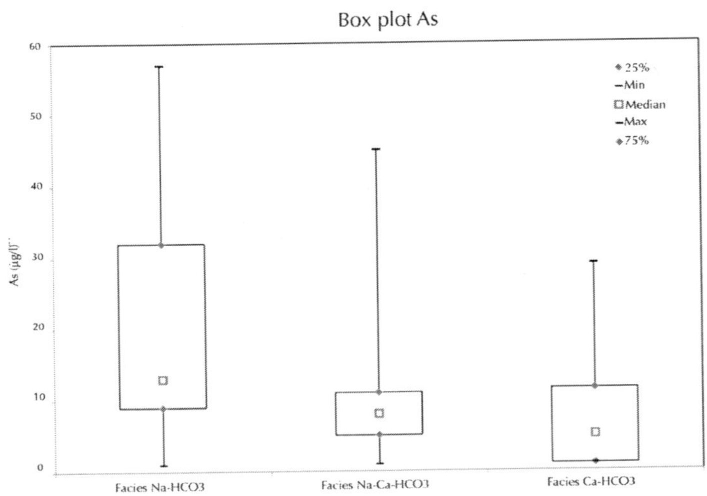

Figure 5: Box plot of mean, median, maximum and minimum values of arsenic distributions versus hydrochemical facies.

Arsenic Speciation and Geochemical Modeling

The two forms of arsenic, arsenate and arsenite are commonly found in ground water (Masscheleyn et al. 1991). Under oxidising conditions, $H_2AsO_4^-$ is dominant at low pH, while at higher pH, $HAsO_4^{2-}$and AsO_4^{3-} becomes dominant. Under reducing conditions, pH up to 9.2, As(III) species H_3AsO_3 is predominant, while $H_2AsO_3^-$ is predominant from pH 9.2 to 12 (Welch and Stollenwerk 2003). The Eh-pH diagram of arsenic species shows that arsenate As (V) is the dominant As species, as $H_2AsO_4^-$and $HAsO_4^{2-}$, in groundwater samples (Figure 6). Similarly, PHREEQC speciation modeling revealed that As (V) species in the groundwater samples are predominantly $H_2AsO_4^-$ and $HAsO_4^{2-}$. Redox potential (Eh) and pH are the most important factors controlling As speciation. Many of the redox processes occur at mineral surfaces and are associated with adsorption/desorption processes. A number of studies have demonstrated that arsenic species are strongly sorbed by oxide and hydroxide minerals, especially iron, aluminum, and manganese oxyhydroxides. The adsorption depends on pH and other solution properties (Arai et al. 2001; Stollenwerk 2003). Positive levels of Eh (> +2 mV) were measured in most of the groundwater samples highlighting the oxidizing conditions. (Table 1). The measured Eh values were converted to *pe* to identify the possible redox sensitive species present in the studied water samples using the following expression:

$$pe = \mathrm{F\ Eh/2.303\ R\ T} \tag{1}$$

Where F is the Faraday constant (F = 96.490 KJ ve - eq) and T is the temperature in K. The pe values of water samples ranges from 3.9 to 11.4 with min and max values, respectively, however most of the samples fall in the range of 5 to 7 indicating the iron and manganese reduction (Table 1).

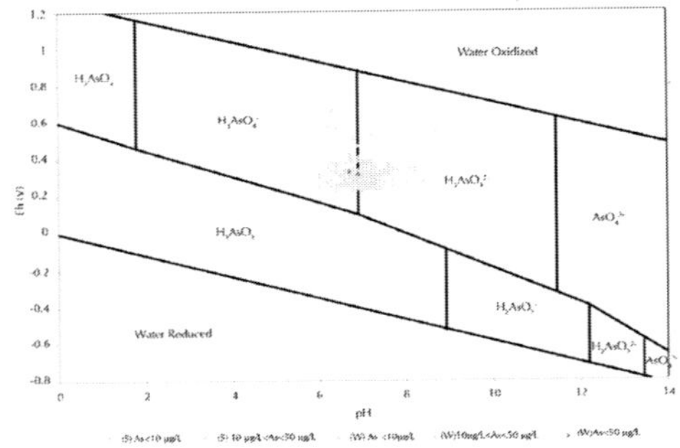

Figure 6: Eh-pH diagram of the arsenic species in the system As–O–H.

Mineralogy (i.e. water-rock interaction) can control the adsorption, desorption and transport of arsenic and other trace elements, and plays an important role in the mechanism of environmental contamination. Thus, geochemical modeling technique was applied to model groundwater evolution by dissolution and precipitation of mineral phases and its relation to the mobilization of As in groundwater. Calculated saturation indexes for selected minerals are presented in Table 3, which may suggest different mineral–solution interactions, precipitation–dissolution and adsorption–desorption processes. All sampled waters are strongly undersaturated with respect arsenolite, claudetite and fluorite oxides and siderite minerals indicating that As should generally remain dissolved after mobilization. Besides, groundwater samples are also undersaturated with respect to carbonate (i.e calcite and dolomite) and sulphate (i.e. gypsum and anhydrite) minerals. Undersaturation with respect to carbonate and arsenic bearing minerals suggest that the groundwater has short residence time and natural equilibrium with these minerals is not reached. On the contrary, most of the groundwater is supersaturated with respect to Al hydroxides such as diaspore, boehmite and gibbsite (Table 3). Groundwater samples show also supersaturation with respect to ferric oxides (hematite, magnetite, maghemite, magnesioferrite, hercynite) and hydroxides (goethite, lepidocrocite, ferrihydrite), suggesting that both Al and Fe mineral phases are probably potential As adsorbents. The results show that Fe-oxides and -oxyhydroxides can precipitate, providing sites for adsorption, according to the following reaction:

$$2Fe^{2+} + 1/2O_2 + 3H_2O = 2FeOOH + 4H^+$$

Table 3: Saturation index values of various mineral phases

	Samples	MJ1	MJ2	MB1	MB2	MB3	MB4	MB5
Saturation Index Values of Minerals	Arsenolite (As_2O_3)	−47.95	−55.92	−57.39	−50.39	−51.62	−55.43	−49.84
	Claudetite (As_2O_3)	−47.66	−55.64	−57.11	−50.11	−51.34	−55.14	−49.56
	Boehmite (γ-AlOOH)	−1.45	0.61	0.56	−0.71	−1.10	0.14	−1.76
	Diaspore (α-AlOOH)	0.31	2.40	2.34	1.07	0.68	1.90	0.02
	Goethite (FeOOH)	1.96	3.86	5.96	4.07	3.85	5.24	2.45
	Lepidocrocite (γ-FeOOH)	1.31	3.32	5.39	3.51	3.29	4.60	1.90
	Magnetite (Fe_3O_4)	6.53	11.68	17.86	12.70	11.94	15.87	7.95
	Gibbsite α-$Al(OH)_3$	−1.08	1.02	0.96	−0.31	−0.70	0.52	−1.35
	Hematite (Fe_2O_3)	6.28	10.08	14.28	10.50	10.06	12.84	7.27
	Siderite ($FeCO_3$)	−4.64	−3.27	−1.55	−2.43	−3.04	−1.49	−3.93
	Magnesioferrite $Mg(Fe^{3+})_2O_4$	−3.13	2.33	6.61	1.60	1.23	4.95	−2.26
	Maghemite (γ-Fe_2O_3)	2.99	−1.03	7.13	3.37	2.93	5.55	0.16
	Fluorite (CaF_2)	−0.89	−1.86	−1.87	−0.15	−0.03	−0.38	−0.04
	Ferrihydrite (~FeOOH)	−0.79	1.09	3.19	1.30	1.09	2.48	−0.31
	Hercynite ($Fe^{2+}Al_2O_4$)	−3.56	1.91	3.78	−0.13	−1.23	2.40	−3.74

Sappa *et al.*

Sappa *et al.* *SpringerPlus* **2014** **3**:237, doi: 10.1186/2193-1801-3-237

The adsorption of arsenic is strong at acidic to neutral conditions, however, increases in pH will result in desorption of arsenic from oxide surfaces and a resultant increase in dissolved concentrations (Fuller et al. 1993). These processes are considered to have been responsible for the release of arsenic in oxidizing aquifers, whereas under reducing condition the reductive dissolution takes place (Dzombak and Morel 1990). As a result of the pH dependence of arsenic adsorption, changes in ground-water pH can promote adsorption or desorption of arsenic. According to calculated saturation indexes it was considered two different mechanism that can lead to the release of arsenic in groundwater (a) the first is the development of high pH, and hence mineral weathering, leading the desorption of adsorbed arsenic (i.e. arsenate species) from natural mineral oxides, or prevents arsenic from being adsorbed in the first place (b) the second is the direct dissolution of mineral phases, derived from volcanic materials, presented in the aquifer system.

Principal Component Analyses (PCA)

Principal component analysis (PCA) is a kind of factor analysis, which is useful to reduce the number of variables in a data set to a few components or factors, that represent most of the variation in the original data simplifying multiple variable interpretation (Hair et al. 1998). These methods have been widely used to identify geochemical controls on the groundwater composition (Seyhan et al. 1985; Join et al. 1997; Hernandez et al. 1991). PCA may result from the correlation of sets of variables representing the same geological origin and/or geochemical source. The PCA was based on the eigenanalysis of the correlation matrix, and hence the Varimax rotation was adopted to maximize the variation explained by the components (Meglin 1991; Reyment and Jvreskog 1996; Everitt et al.2011). All PCs are uncorrelated (i.e. orthogonal) to one another. Eigenvalues describe the amount of variance explained by each PC, and thus decrease with each successive PC extracted. The number of significant principal components is selected on the basis of the Kaiser criterion and only factors with eigenvalues greater than or equal to 1 were considered (Kaiser 1960). Eigenvectors (or PC loadings) indicate the relative contribution that each element makes to that PC score (Webster 2001). In the present study, two data sets of the selected physico-chemical parameters and trace elements were used for PCA to identify the main hydrogeochemical processes governing the groundwater chemistry and to find the effects of some chemical components on arsenic contamination. The first data set for PCA includes EC, TDS, pH, T, Eh, As and major ions (Ca, Mg, Na, K, HCO_3, Cl, SO_4, NO_3, V and F) of the 231 water samples from wells and springs measured in 2007 to 2009. A second PCA was applied with the objective to identify the relationship among the trace elements and their origin. Principal component analysis (PCA) was performed using XLSTAT. The correlation between the arsenic and other trace element concentrations of groundwater was obtained in the form of Pearson correlation coefficients to find out relationships between variables and the participation of individual chemical parameters in several influence factors.

Table 4 summarizes the first PCA results on the data matrices of spring and groundwater samples including the loadings of each PC, percentage of variance and cumulative percentage of variance of each factor and their respective eigenvalues. The results revealed that the contribution of the first five principal components accounts for approximately 79.4% of the total variance in groundwater data. The first component (PC1), explaining 41.3% of the total variance, exhibits negative loading on As concentration, strong positive loadings on EC, Na, Mg, Ca, Cl, HCO_3, SO_4 and TDS, moderate loadings on Temperature and weak loadings on pH, Eh, K, F, NO_3 and V. PC2 explains for 15.8% of the variance and represented by K, F, As and V. PC3 accounts for 8.1% of the variance and show moderately positive loading only on NO_3 suggesting the source of pollution is possibly related to agricultural activities. PC4 and PC5 are responsible for 7.5% and 6.5% of the total variance and show moderate positive loadings for As, T, Ph and Eh, respectively. Four principal components were extracted on the spring water data matrix explaining 79.5% of the total variance. The first two PCs explain 45.4% and 18.9% of the variance, respectively, and account for the majority of the variance in the original dataset. The first one is mainly participated by EC, Na, Mg, Ca, Cl, NO_3, HCO_3, SO_4 and TDS, while the second is characterized by Eh, F, As and V. PC3, which accounts for 8.5% of the variance, shows moderate positive loadings for As and pH. The fourth component (PC4) show weak positive loadings for all components. Figure 7A and B shows the loading plot for the first two PCs (PC1 and PC2) identifying different groups in springs and groundwater, respectively. Evaluation of the PC loadings (for springs and groundwater data matrices) show that most of the physicochemical parameters with greatest positive PC1 loadings typically occurred in groundwater that has flowed through volcanic materials. Pearson correlation matrix of springs and groundwater samples show that the correlation coefficients between arsenic and physicochemical components are very low ($r < 0.5$), and are statistically insignificant.

Table 4: Principal component loadings of physico-chemical parameters for springs and wells including variance % and cumulative % and their respective eigenvalues

	Springs				Wells				
	PC1	PC2	PC3	PC4	PC1	PC2	PC3	PC4	PC5
T	0.199	0.225	−0.088	0.305	0.071	−0.089	0.320	0.518	−0.405
pH	0.039	−0.011	0.699	−0.279	−0.051	−0.262	0.126	0.281	0.617
25°C	0.364	−0.028	−0.006	−0.115	0.383	0.006	−0.062	−0.010	0.045
ENHE	0.028	0.296	0.152	−0.271	−0.037	0.255	0.055	0.168	0.600
Na	0.324	0.067	0.041	0.249	0.334	0.005	0.262	0.098	−0.033
K	0.096	0.475	−0.144	−0.009	0.117	0.511	−0.219	0.069	0.029
Mg	0.271	0.120	−0.249	−0.392	0.355	0.050	0.000	0.042	−0.079
Ca	0.331	−0.171	0.020	−0.188	0.349	−0.083	−0.177	−0.053	0.065
Cl	0.311	−0.176	−0.024	0.046	0.296	−0.180	0.366	−0.022	−0.044

NO3	0.305	−0.088	0.118	0.405	0.145	−0.082	0.590	−0.334	0.163
HCO3-	0.335	0.011	−0.053	−0.353	0.352	0.069	−0.253	0.025	0.040
F-	0.062	0.487	0.111	0.120	0.013	0.491	0.294	0.001	0.120
SO42-	0.302	−0.082	0.180	0.397	0.298	−0.011	−0.124	0.136	0.083
As	−0.029	0.361	0.490	0.032	−0.085	0.346	0.152	0.537	−0.121
V	0.009	0.419	−0.306	0.036	0.003	0.426	0.194	−0.432	−0.114
TDS	0.362	−0.018	−0.014	−0.161	0.378	0.040	−0.134	0.009	0.047
Eigenvalues	7.27	3.03	1.37	1.03	6.61	2.54	1.30	1.20	1.04
Variance %	45.4	18.9	8.5	6.4	41.3	15.8	8.1	7.5	6.5
Cumulative %	45.4	64.4	73.0	79.5	41.3	57.2	65.3	72.9	79.4

Sappa *et al.*

Sappa *et al. SpringerPlus* 2014 3:237, doi: 10.1186/2193-1801-3-237

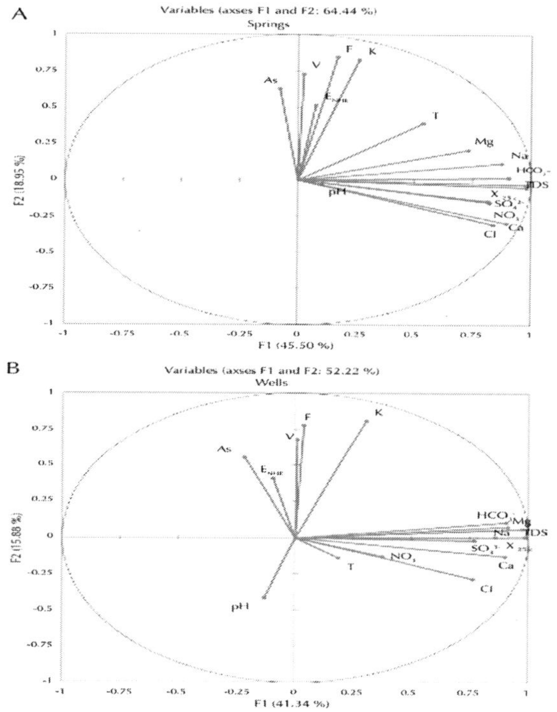

Figure 7: Scatter plot between PC1 and PC2 for physico-chemical parameters of (A) springs and (B) wells.

Principle Component Analysis (PCA) were applied for trace element data of groundwater samples from Monte Jugo and Monte Bisenzio areas to identify similarities and dissimilarities in hydrogeochemical properties and to make predictions about the As mobilization. For the groundwater samples from Monte Jugo area, the application of principal components analysis generated 7 orthogonal principal components and explained 100% of the total variation. The principal component analysis of standardized parameters resulted in seven components accounting for 60.9%, 22.6%, 9.56%, 4.1%, 1.18%, 0.85% and 0.67% of the total variance, respectively. The variables that participated in PCA as well as their obtained loadings and eigenvalues are shown in Table 5. The first component show moderately to strong positive loadings for Li, B, Co, As, Se, Sr, Mo, and U, while the second component has significant positive loadings for Li, V, Co, Rb, and moderate loadings for Sr and Mo. This indicate that these components accounted for the maximum variance of the PCA and was representative of arsenic and these trace elements release due to groundwater circulation in the volcanic materials. The third component (PC3) show weak to moderate positive loadings for Be, V, Co, Ni, As Al and Mn suggesting that the As mobilization is probably controlled by adsorption/desorption processes on the oxide/hydroxide minerals. Figure 8 shows the score plot for the first two PCs (PC1 and PC2), explaining 83.61% of the total variance within all the measured parameters. The loading plot between PC1 and PC2 has four distinct groups associated with (A) Co, Li, Mo, B, As and Sr, (B) Rb, V, Mn, Al, Be, Ni, Cu, Zn, Cr, and Fe, (C) Sb elements, and (D) U and Se (Figure 6). Table 6 presents the correlation matrix of the 19 trace element variables. Only those, with correlation values higher than 0.5, were considered. Arsenic is strongly correlated with (Pearson $r > 0.50$) the following trace elements: Sr ($r = 0.91$), Li ($r = 0.84$), B ($r = 0.89$), Co ($r = 0.73$), U ($r = 0.77$). Mo ($r = 0.71$), and Se ($r = 0.6$). High correlation coefficients between As and the mentioned trace elements show that these trace elements in groundwater had similar hydrochemical characteristics in the study area. It is very well known that volcanic rocks and sediments derived from them are also associated with elevated arsenic levels in ground water. The high concentrations of molybdenum, uranium, lithium, and boron are predominantly associated with volcanic materials and/or interaction of groundwater with geothermal waters (Aiuppa et al. 2003; Vivona et al. 2007; Arnórsson and Óskarsson 2007). The results suggest that volcanic materials are a significant source of As and these trace elements in groundwater. Groundwater, relatively more oxidizing, was characterized by greater concentrations of trace elements such as Mo, Se, B, As, and U etc. These trace elements having greatest positive PC1 loadings exhibited more soluble in oxidizing environments. According to, PHREEQC speciation modeling Mo predominant as MoO_4^{2}, selenium mostly occurred as $HSeO_3^-$ and SeO_3^{2-} species and boron was composed of $B(OH)_3$ and/or $B(OH)_4^-$ species. Uranium exists in U(IV) or U(VI) oxidation state in groundwater depending on the environmental conditions. The dominant dissolved species in the oxidized groundwater were uranyl carbonate complexes ($UO_2CO_2^{2-}$; $UO_2CO_3^{4-}$). In fact, the concentrations of these elements in groundwater may be due to the high dissolved oxygen content in groundwater. Table 7 summarized the PCA results of groundwater samples from Monte Bisenzio area including the loadings of each PC and their respective eigenvalues. The application of principal components analysis, generated 18 orthogonal principal components and the first seven explained 98,03% of the total variation. The first PC is responsible for 51.3% of the total variance and show strong positive loadings for Li, B, Be, Co, Rb, Sr and U. PC2 show positive loadings for all parameters except for Cr and Se and is best represented by Mn, Fe, Zn, As and V. The first two PCs (PC1 and PC2) explained 76.2% of the total variance within all the measured parameters. The loading plot between PC1 and PC2 includes four distinct groups. The identified groups were associated with the following elements: (A) Mn, Al, Cu, Ni, Be, Li, Sr, Co, Rb and U, (B) As, Mo, Fe, Zn, V and Sb, (C) Cr and (D) Se elements (Figure 9). According to Pearson correlation matrix, arsenic show positive correlations with Mn (r = 0.53), Fe (r = 0.56), and Mo (r = 0.58) whereas, correlates negatively with Cr, Co, Ni, Cu, Se, Rb, Sr, and U (Table 8). The positive correlation with Fe and Mn suggests common geogenic origin of these elements providing the presence of Fe/Mn oxyhydroxides could lead to desorption of arsenic. On the contrary, strong correlation coefficients among the other trace elements (i.e. Li, Sr, Rb, U, B, Co, Be and Co) and their close relation indicated by PCA and bivariate plots suggest the origin of these elements probably related to the volcanic source aquifer materials. The third component (PC3), explaining 7.7 % of the variance, exhibits moderate positive loadings on Ni and Se, while PC4 (accounts for 5.4 of the variance) show moderately positive loading only for As.

Table 5: Variable loadings to the first seven PCs for the groundwater samples from Monte Jugo area

Variable	PC1	PC2	PC3	PC4	PC5	PC6	PC7
Li	0.541	0.833	−0.012	0.039	−0.013	0.079	−0.077
Be	−0.771	0.242	−0.083	0.555	−0.045	−0.176	0.002
B	0.779	0.476	0.201	0.320	0.095	0.085	0.082
Al	−0.752	0.267	0.588	−0.105	−0.029	−0.067	0.027
V	−0.425	0.778	0.146	−0.408	−0.124	−0.094	0.053
Cr	−0.878	0.194	−0.362	−0.243	0.003	−0.046	0.010
Mn	−0.697	0.346	0.603	0.042	−0.085	0.144	0.026
Fe	−0.979	0.117	−0.148	−0.019	−0.029	0.002	0.073
Co	0.334	0.934	0.017	−0.029	0.063	−0.082	−0.060
Ni	−0.926	0.170	0.039	−0.006	0.322	0.090	0.020

Cu	−0.972	0.141	−0.135	−0.037	0.101	0.066	−0.028
Zn	−0.927	0.111	−0.323	−0.098	0.114	0.036	0.029
As	0.838	0.484	0.104	−0.106	−0.013	0.080	0.188
Se	0.921	−0.208	−0.122	−0.253	−0.012	0.087	−0.145
Rb	−0.480	0.835	−0.161	0.116	−0.052	0.085	−0.154
Sr	0.802	0.561	−0.181	0.088	−0.031	−0.003	−0.024
Mo	0.617	0.504	−0.583	−0.073	0.065	−0.095	0.084
Sb	−0.760	−0.031	−0.565	0.127	−0.221	0.176	0.075
U	0.982	−0.092	−0.132	0.028	0.032	0.025	0.087
Eigenvalues	11.575	4.311	1.817	0.780	0.224	0.163	0.129
Variance %	60.920	22.691	9.565	4.107	1.181	0.856	0.679
Cumulative %	60.9	83.6	93.1	97.2	98.4	99.3	100

Sappa *et al.*

Sappa *et al. SpringerPlus* **2014 3**:237, doi: 10.1186/2193-1801-3-237

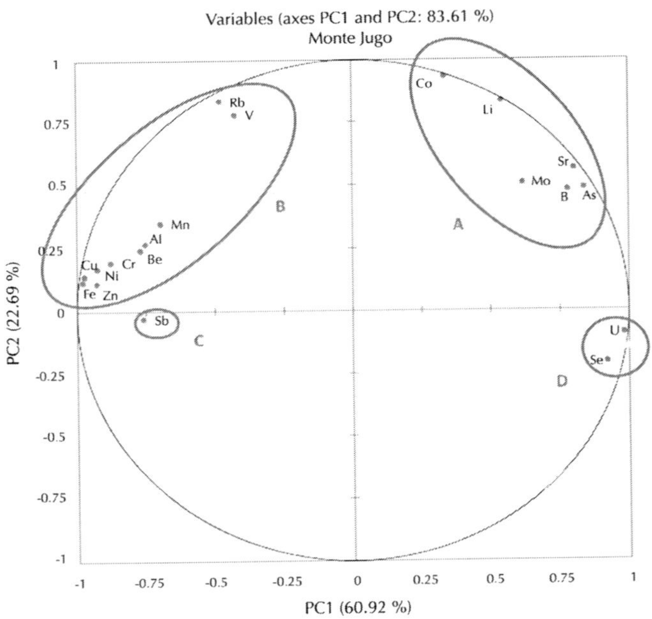

Figure 8: A two dimensional sample score plot for the groundwater samples from Monte Jugo area (total PC variance 83.61%).

Table 6: Pearson's correlation matrix for trace elements measured in groundwater samples from Monte Jugo area

	Li	Be	B	Al	V	Cr	Mn	Fe	Co	Ni	Cu	Zn	As	Se	Rb	Sr	Mo	Sb	U
Li	1	−0.206	**0.827**	−0.203	0.390	−0.323	−0.085	−0.437	**0.955**	−0.359	−0.403	−0.410	**0.843**	0.335	0.461	**0.909**	**0.743**	−0.415	0.452
Be		1	−0.343	0.550	0.300	0.627	0.573	0.785	−0.037	**0.718**	**0.758**	0.702	−0.609	**−0.906**	0.637	−0.416	−0.330	0.675	**−0.758**
B			1	−0.381	−0.078	**−0.745**	−0.238	**−0.740**	0.694	−0.595	**−0.717**	**−0.750**	**0.891**	0.508	0.018	**0.879**	0.585	−0.680	**0.716**
Al				1	0.667	0.528	**0.961**	0.685	0.013	**0.750**	0.685	0.542	−0.429	**−0.803**	0.468	−0.569	−0.658	0.214	**−0.844**
V					1	0.575	0.634	0.501	0.595	0.487	0.498	0.457	0.082	−0.483	**0.773**	0.035	0.080	0.180	−0.522
Cr						1	0.444	**0.940**	−0.108	**0.830**	**0.936**	**0.974**	−0.655	**−0.749**	0.608	−0.552	−0.209	**0.827**	**−0.839**
Mn							1	0.637	0.081	**0.714**	0.643	0.482	−0.341	**−0.789**	0.543	−0.469	−0.627	0.230	**−0.792**
Fe								1	−0.226	**0.913**	**0.984**	**0.969**	**−0.763**	**−0.913**	0.579	−0.695	−0.453	**0.834**	**−0.947**
Co									1	−0.138	−0.192	−0.206	**0.718**	0.119	0.613	**0.786**	0.676	−0.329	0.233
Ni										1	**0.957**	**0.905**	−0.682	**−0.891**	0.567	−0.666	−0.493	0.622	**−0.916**
Cu											1	**0.977**	**−0.758**	**−0.891**	0.606	−0.682	−0.449	**0.794**	**−0.948**
Zn												1	**−0.739**	**−0.815**	0.570	−0.636	−0.314	**0.854**	**−0.873**
As													1	0.665	−0.049	**0.911**	**0.715**	−0.693	**0.779**
Se														1	−0.595	0.626	0.531	−0.650	**0.922**
Rb															1	0.128	0.186	0.459	−0.537
Sr																1	**0.873**	−0.509	**0.759**
Mo																	1	−0.189	0.640
Sb																		1	−0.662
U																			1

Values in bold are different from 0 with a significance level alpha = 0.05.

Sappa *et al.*

Sappa *et al.* *SpringerPlus* 2014 **3**:237, doi: 10.1186/2193-1801-3-237

Table 7: Variable loadings to the first seven PCs for the groundwater samples from Monte Bisenzio area

Variable	PC1	PC2	PC3	PC4	PC5	PC6	PC7
Li	0.959	0.243	0.112	0.034	−0.014	0.013	−0.011
Be	0.848	0.285	−0.093	0.306	−0.163	−0.074	−0.089
B	0.966	0.147	−0.006	0.148	−0.071	−0.016	−0.103
Al	0.380	0.486	−0.511	0.052	0.335	0.402	0.279
V	−0.818	0.512	0.034	−0.150	0.029	−0.053	0.125
Cr	−0.559	−0.685	−0.036	0.291	0.305	0.113	−0.112
Mn	0.442	0.843	0.175	0.103	0.125	0.001	−0.067
Fe	−0.551	0.797	0.101	−0.053	−0.029	−0.024	0.002
Co	0.974	0.166	0.110	−0.032	−0.068	0.014	0.020
Ni	0.455	0.061	0.617	−0.380	0.299	0.356	−0.202
Cu	0.541	0.213	−0.441	−0.093	0.535	−0.382	−0.138
Zn	−0.575	0.764	0.101	−0.167	−0.061	−0.044	0.064
As	−0.171	0.622	0.137	0.717	0.005	0.143	−0.068
Se	0.216	−0.375	0.709	0.254	0.297	−0.266	0.290
Rb	0.966	0.111	0.094	−0.125	−0.104	0.002	0.109
Sr	0.979	0.175	0.026	−0.063	−0.024	−0.040	0.023
Mo	−0.436	0.883	0.080	−0.035	0.029	−0.093	−0.008
Sb	−0.827	0.478	0.090	−0.029	0.080	−0.105	−0.109
U	0.966	0.127	−0.071	−0.125	−0.039	−0.087	0.040
Eigenvalues	9.7	4.7	1.4	1.0	0.74	0.57	0.30
Variance %	51.3	24.8	7.7	5.4	3.9	3.0	1.6
Cumulative %	51.3	76.2	83.9	89.4	93.3	96.4	98.0

Sappa *et al.*

Sappa *et al. SpringerPlus* 2014 **3**:237, doi: 10.1186/2193-1801-3-237

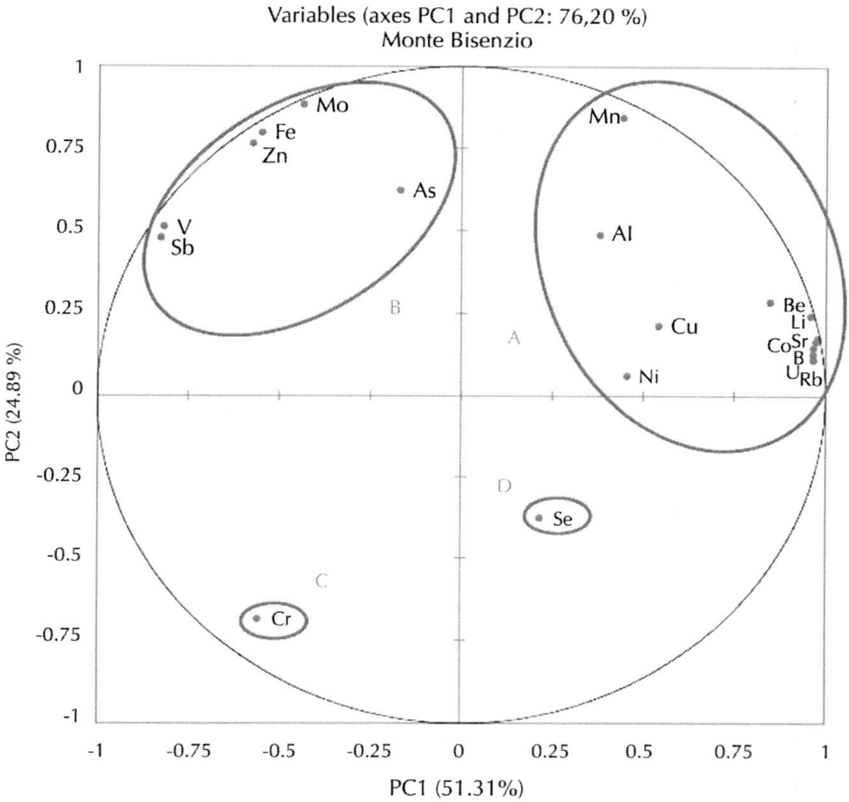

Figure 9: A two dimensional sample score plot for the groundwater samples from Monte Bisenzio area (total PC variance 76.20 %).

Table 8: Pearson's correlation matrix for trace elements measured in groundwater samples from Monte Bisenzio area

	Li	Be	B	Al	V	Cr	Mn	Fe	Co	Ni	Cu	Zn	As	Se	Rb	Sr	Mo	Sb	U
Li	1	**0.877**	**0.967**	0.423	**−0.672**	**−0.695**	**0.662**	−0.327	0.990	**0.504**	**0.507**	−0.356	0.026	0.191	**0.961**	**0.980**	−0.194	**−0.674**	**0.934**
Be		1	**0.932**	0.428	**−0.593**	**−0.617**	**0.604**	−0.268	0.850	0.185	**0.475**	−0.316	0.213	0.043	**0.799**	**0.862**	−0.135	**−0.553**	**0.843**
B			1	0.393	**−0.754**	**−0.605**	**0.559**	−0.430	0.961	0.382	**0.522**	**−0.463**	0.032	0.142	**0.924**	**0.963**	−0.291	**−0.719**	**0.936**
Al				1	−0.063	−0.394	**0.513**	0.097	0.379	0.056	**0.513**	0.077	0.240	−0.374	0.359	0.423	0.192	−0.164	0.424
V					1	0.051	0.041	**0.877**	**−0.706**	−0.291	−0.319	**0.878**	0.343	−0.323	**−0.701**	**−0.690**	**0.813**	**0.939**	**−0.684**
Cr						1	**−0.745**	−0.267	**−0.692**	−0.279	−0.331	−0.284	−0.117	0.215	**−0.703**	**−0.699**	−0.367	0.146	**−0.689**
Mn							1	0.442	**0.582**	0.360	0.404	0.380	**0.536**	−0.056	**0.501**	**0.569**	**0.578**	0.043	**0.483**
Fe								1	−0.390	−0.137	−0.168	**0.921**	**0.566**	−0.365	−0.418	−0.392	**0.947**	**0.815**	−0.428
Co									1	**0.509**	**0.472**	−0.414	−0.068	0.197	**0.985**	**0.990**	−0.270	**−0.724**	**0.956**
Ni										1	0.075	−0.137	−0.157	0.354	**0.499**	**0.470**	−0.110	−0.263	0.407
Cu											1	−0.203	−0.118	−0.082	0.451	**0.559**	−0.034	−0.293	**0.598**
Zn												1	0.448	−0.365	−0.432	−0.417	**0.946**	**0.850**	−0.449
As													1	−0.050	−0.170	−0.105	**0.585**	0.422	−0.194
Se														1	0.199	0.157	−0.346	−0.276	0.103
Rb															1	**0.981**	−0.321	**−0.759**	**0.965**
Sr																1	−0.264	**−0.714**	**0.985**
Mo																	1	**0.801**	−0.307
Sb																		1	**−0.722**
U																			1

Values in bold are different from 0 with a significance level alpha = 0.05.

Sappa *et al.*

Sappa *et al. SpringerPlus* 2014 3:237, doi: 10.1186/2193-1801-3-237

CONCLUSIONS

This paper provides information on the occurrence and the distribution of arsenic and other trace elements in the most important water supply networks of Viterbo area, Central Italy. To find out the major factors and geochemical processes affecting major and trace element concentrations in groundwater a combined geochemical modeling and principal component analysis techniques were applied. Based on the dominance of major anions and cations of water samples three hydrochemical facies identified: (i) Facies A- Na-HCO_3 type water, (ii) Facies B- no dominant type, Na-Ca-HCO_3 and (iii) Facies C- Ca-HCO_3. According to speciation modeling, the dominant As species in the waters is arsenate As (V) in the forms of $HAsO_4^{2-}$ and $H_2AsO_4^-$ reflecting oxidizing conditions. Geochemical modelling show that udersaturation with respect to Arsenic-bearing phases in the groundwater doesn't contribute the arsenic mobilization. However, most of the groundwater samples show supersaturation with respect to Fe and Al oxide and hydroxide mineral phases suggesting that As concentrations in groundwater probably was controlled by adsorption/desorption processes on these minerals. Principal component analysis reveals the similarities in the concentrations of trace elements in the water samples resulting from different geochemical processes. PCA results show high positive loadings for arsenic and other trace elements such as Li, B, Rb, Co, Mo, U and Sr suggesting the primary sources of these elements are probably derived from weathering and/or dissolution of volcanic materials in oxidizing conditions. Besides, positive loadings are also observed for As, Fe and Al confirming the results of geochemical modeling that the adsorption/desorption process as a possible mechanism of As release in groundwater. The high correlation coefficients between the mentioned elements and their close relation indicated by the PCA seem to be consistent with the hypothesis that the past volcanic activity and related volcanic materials may have been a significant source of these elements.

The results show that the knowledge on geochemistry of major and trace elements and the mechanisms of their release into groundwater is important to the development effective strategies for appropriate remediation techniques minimizing elevated levels of naturally occurring contaminants. However, according to preliminary results, further investigations should be carried out concerning (i) a more detailed study on the geochemistry of aquifer rocks and sediments to confirm the volcanic materials as the source of As and other trace elements and (ii) studies of varies isotopes to develop a hydrogeochemical model and to systematize the hydrologic cycles.

AUTHORS' CONTRIBUTIONS

GS planned the measurement campaign and research work, coordinating the writing of the manuscript. The interpretation of data analysis and drafting of manuscript has been done by SE. FF participated in the design of the study and performed the statistical analysis. All authors read and approved the final manuscript.

ACKNOWLEDGEMENTS

The authors gratefully acknowledge the support provided by Italian National Institute of Health (ISS).

REFERENCES

1. Aiuppa A, Bellomo S, Brusca L, D'Alessandro W, Federico C (2003) Natural and anthropogenic factors affecting groundwater quality of an active volcano (Mt. Etna, Italy). Appl Geochem 18:863-882
2. Aiuppa A, Federico C, Allard P, Gurrieri S, Valenza M (2005) Trace metal modelling of groundwater–gas–rock interactions in a volcanic aquifer: Mount Vesuvius, Southern Italy. Chem Geol 2163–4:289-311
3. Anawar HM, Akai J, Komaki K, Terao H, Yoshioka T, Ishizuka T, Safiullah S, Kato K (2003) Geochemical occurrence of arsenic in groundwater of Bangladesh: sources and mobilization processes. J Geochem Explor 77:109-131
4. Appelo CAJ, Postma D (1993) Geochemistry. Balkema, Rotterdam: Groundwater and Pollution.

5. Arai Y, Elzinga EJ, Sparks DL (2001) X-ray Absorption Spectroscopic Investigation of Arsenite and Arsenate Adsorption at the Aluminum Oxide-Water Interface. Journal of Colloid Interface Science 235(1):80-88
6. Arnórsson S, Óskarsson N (2007) Molybdenum and tungsten in volcanic rocks and in surface and <100°C ground waters in Iceland. Geochim Cosmochim Acta 71:284-304
7. Baiocchi A, Lotti F, Piscopo V, Chiocchini U, Madonna S, Manna F (2007) Hydraulic interactions between aquifers in Viterbo area (Central Italy). In: Howard KWF (ed) Urban Groundwater -Meeting the Challenge, vol 8. London: Taylor & Francis Group, pp 223-238 vol 8. IAH Selected Papers
8. Baiocchi A, Lotti F, Piscopo V (2012) Conceptual Hydrogeological Model and Groundwater Resource Estimation in a Complex Hydrothermal Area: The Case of the Viterbo Geothermal Area (Central Italy). Journal of Water Resource and Protection 4:231-247
9. Ballantyne JM, Moore JN (1988) Arsenic geochemistry in geothermal systems. Geochim Cosmochim Acta 52:475-48313 Webster and Nordstrom 2003
10. Barberi F, Buonasorte G, Cioni R, Fiordelisi A, Foresi L, Iaccarino S, Laurenzi MA, Sbrana A, Vernia L, Villa IM (1994) Plio-Pleistocene geological evolution of the geothermal area of Tuscany and Latium. Mem Descr Carta Geol Ital 49:77-134
11. Barbieri M, Battistel M, Garone A (2013) The geochemical evolution and management of a coastal wetland system: A case study of the Palo Laziale protected area. J Geochem Explor 126–127:67-77
12. Boni C, Bono P, Capelli G (1986) Schema idrogeologico dell'Italia centrale. Memorie Società Geologica Italiana 35:991-1012
13. Borghetti G, La Torre P, Sbrana A, Sollevanti F (1983) Geothermal exploration in Monti Cimini permit (North Latium, Jtaly). European Geothermal Update. 419-432
14. Bose P, Sharma A (2002) Role of iron in controlling speciation and mobilization of arsenic in subsurface environment. Water Res 36:4916-4926

15. Bratus A, Caputo DR, De Nardi R, Santarato G (2006) Geophysical exploration integrated with hydrogeological data of a deep multiaquifer system: results from a pilot site. IAGA WG 1.2 on Electromagnetic Induction in the Earth Extended Abstract 18th Workshop El Vendrell, September 17-23. Spain.
16. Buschmann J, Berg M, Stengel C, Sampson ML (2007) Arsenic and manganese contamination of drinking water resources in Cambodia: coincidence of risk areas with low relief topography. Environ Sci Technol 41(7):2146-2152
17. Capelli G, Mazza R, Gazzetti C (2005) Strumenti e strategie per la tutela e l'uso compatibile della risorsa idrica nel Lazio. Gli acquiferi vulcanici. Bologna: Pitagora Editrice. 191 pp
18. Casentini B, Pettine M (2010) Effects of desferrioxamine-B on the release of arsenic from volcanic rocks. Appl Geochem 25(11):1688-1698
19. Casentini B, Pettine M, Millero F (2010) Release of Arsenic from Volcanic Rocks through Interactions with Inorganic Anions and Organic Ligands. Aquat Geochem 16(3):373-393
20. Chapman D (1992) Water Quality Assessment; A Guide to the Use of Biota, Sediments and Water in Environmental Monitoring. Cambridge: Chapman & Hall University Press.
21. Charlet L, Polya DA (2006) Arsenic in shallow, reducing groundwaters in southern Asia: an environmental health disaster. Elements 2:91-96
22. Chenini I, Khmiri S (2009) Evaluation of ground water quality using multiple linear regression and structural equation modeling. International Journal of Environmental Science and Technology 6(3):509-519
23. Chiocchini U, Castaldi F, Barbieri M, Eulilli V (2010) A stratigraphic and geophysical approach to studying the deepcirculating groundwater and thermal springs, and their recharge areas, in Cimini Mountains–Viterbo area, central Italy. Hydrogeol J 18:1319-1341
24. Cimarelli C, De Rita D (2006) Structural evolution of the Pleistocene Cimini trachytic volcanic complex (Central Italy). Bull Volcanol 68:538-548

25. Coli M, Principi G, Peccerillo A (1991) Evoluzione geodinamica recente dell'Appennino settentrionale e attività magmatica tosco–laziale: vincoli e problemi [Recent geodynamic evolution of the northern Apennines and Tuscany–Latium magmatic activities: constraints and problems. Studi Geol Camerti 1:403-413

26. Dall'Aglio M, Duchi V, Minissale A, Guerrini A, Tremori M (1994) Hydrogeochemistry of the volcanic district in the Tolfa and Sabatini Mts. in central Italy. J Hydrol 154:195-217

27. De Rita D, Funiciello R, Rossi U, Sposato A (1983) Structure and evolution of the Sacrofano Baccano caldera, Sabatini volcanic complex. J Volcanol Geotherm Res 17:219-236

28. Della Vedova B, Pellis G, Foucher JP, Rehault JP (1984) Geothermal structure of the Tyrrhenian Sea. Mar Geol 55:271-289

29. Drever JI (1997) The Geochemistry of Natural Waters Prentice Hall. New Jersey: Englewood Cliffs.

30. Dzombak DA, Morel FMM (1990) Surface complexation modelling-Hydrous ferric oxide. New York: Wiley & Sons. p 393

31. Everitt BS, Landau S, Leese M, Stahl D (2011) Cluster Analysis. Hoboken, NJ, USA: John Wiley & Sons Ltd.. 330 pp

32. Foster SSD, Lawrence AR, Morris BL (1997) Groundwater in Urban Development: Assessing Management Needs and Formulating Policy Strategies. Washington DC: World Bank Technical Paper.

33. Fuller CF, Davis JA, Waychunas GA (1993) Surface chemistry of ferrihydrite: part 2. Kinetics of arsenate adsorption and coprecipitation. Geochim Cosmochim Acta 57:2271-2282

34. Giammanco S, Ottaviani M, Valenza M, Veschetti E, Principio E, Giammanco G, Pignato S (1998) Major and trace elements geochemistry in the groundwaters of a volcanic area: Mount Etna (Sicily, Italy). Water Res 32:19-30

35. Hair JF, Anderson RE, Tatham RL, Black W (1998) Multivariate Data Analysis. New Jersey: Prentice-Hall. pp 87-138

36. Hernandez MA, Gonzalez N, Levin M (1991) Multivariate analysis of a coastal phreatic aquifer using hydrochemical and isotopic indicators, Buenos Aires, Argentina, Proceedings of the International Association on Water Pollution Research and Control's International Seminar on Pollution, Protection and Control of Groundwater. Water Science Technology 24:139-146

37. Johannesson KH, Tang J (2009) Conservative behavior of arsenic and other oxyanion-forming trace elements in an oxic groundwater flow system. J Hydrol 378:13-28
38. Join JL, Coudray J, Longworth K (1997) Using principal components analysis and Na/Cl ratios to trace groundwater circulation in a volcanic island: the example of reunion. J Hydrol 190(1–2):1-18
39. Kaiser HF (1960) The application of electronic computers to factor analysis. Educ Psychol Meas 20:141-151
40. Kumar P, Kumar M, Ramanathan AL, Tsujimura M (2010) Tracing the factors responsible for arsenic enrichment in groundwater of the middle Gangetic Plain. India: a source identification perspective. Environ Geochem Health 32(2):129-146
41. Langmuir D (1996) Aqueous Environmental Geochemistry Prentice Hall Publishers. New Jersey: Upper Saddle River.
42. López DL, Bundschuh J, Birkle P, Armienta MA, Cumbal L, Sracek O, Cornejo L, Ormachea M (2012) Arsenic in volcanic geothermal fluids of Latin America. Sci Total Environ 429:57-75
43. Masscheleyn PH, Delaune RD, Patrick WH (1991) Effect of redox potential and pH on arsenic speciation and solubility in a contaminated soil. Environmental Science and Technology J 25:1414-1419
44. Meglin RR (1991) Examining large databases: a chemometric approach using principal component analysis. J Chemomet 5:163-179
45. Meharg AA, Rahman MM (2003) Arsenic contamination of Bangladesh paddy field soils: implications for rice contribution to arsenic consumption. Environ Sci Technol 37:229-234
46. Minissale A (2004) Origin, transport and discharge of CO_2 in central Italy. Earth Sci Rev 66:89-141
47. Minissale A, Duchi V (1988) Geothermometry on fluids circulating in a carbonate reservoir in north–central Italy. J Volcanol Geotherm Res 35:237-252
48. Mongelli F, Zito G, Ciaranfi N, Pieri P (1989) Interpretation of heat-flow density of the Apennine chain, Italy. Tectonophysics 164:267-280
49. National Research Council (1999) Arsenic in Drinking Water. Washingon DC: National Academy Press.

50. National Research Council (2001) Arsenic in Drinking Water 2001 Update. Washington DC: National Academy Press.
51. Nickson RT, McArthur JM, Ravenscroft P, Burgess WG, Ahmed KM (2000) Mechanism of arsenic release to groundwater. Bangladesh and West Bengal Appl Geochem 15:403-413
52. Pagano G, Menghini A, Floris S (2000) Ground Water Budget of the Vulsini Basin. Geologia Tecnica & Ambiente 3:31-42
53. Palladino DM, Simei S, Sottili G, Triglia R (2010) Integrated approach for the reconstruction of stratigraphy and geology of Quaternary volcanic terrains: An application to the Vulsini Volcanoes (Central Italy). In: Groppelli G, Viereek-Goette L (eds) Stratigraphy and Geology of Volcanic Areas: Geological Society of America Special Paper, 464, vol 464.63-84
54. Parisi S, Paternoster M, Perri F, Mongelli G (2011) Source and mobility of minor and trace elements in a volcanic aquifer system: Mt. Vulture (Southern Italy). J Geochem Explor. doi:10.1016/j.gexplo.2011.06.010
55. Parkhurst DL, Appello CAJ (1999) User's guide to PHREEQC (version 2) – A computer program for speciation, batch-reaction, one-dimensional transport, and inverse geochemical calculations. Denver, Colorado. p 312
56. Perini G, Conticelli S, Francalanci L, Davidson JP (2000) the relationship between potassic and calc-alcaline post-orogenic magmatism at Vico volcano, central Italy: a case story. J Volcanol Geotherm Res 95:243-268
57. Piper AM (1944) A graphic procedure in the geochemical interpretation of water analysis. American Geophysics Union, Transactions 25:914-923
58. Piscopo V, Barbieri M, Monetti V, Pagano G, Pistoni S, Ruggi E, Stanzione D (2006) Hydrogeology of thermal waters in Viterbo area, central Italy. Hydrogeol J 14:1508-1521
59. Reyment RA, Jvreskog KG (1996) Applied factor analysis in the natural sciences. Cambridge: Cambridge University Press. 383pp
60. Roychowdhury T, Uchino T, Tokunaga H, Ando M (2002) Arsenic and other heavy metals in soils from an arsenic-affected area of West Bengal, India. Chemosphere 49:605-618

61. Seyhan E, van-de-Griend AA, Engelen GB (1985) Multivariate analysis and interpretation of the hydrochemistry of dolomitic reef aquifer, Northern Italy. Water Resour Res 21(7):1010-1024
62. Shakeel AIM, Amal KH (2011) The socioeconomic impact of Arsenic poisoning in Bangladesh. Journal of Toxicology and Environmental Health Sciences 3(3):65-73
63. Smedley PL, Kinniburgh DG (2002) A review of the source, behaviour and distribution of arsenic in natural waters. Appl Geochem 17:517-568
64. Stollenwerk KG (2003) Geochemical processes controlling transport of arsenic. In: Welch AH, Stollenwerk KG (eds) Arsenic in Ground Water: Geochemistry and Occurrence, Boston: Kluwer Academic Publishers. pp 67-100
65. Vivona R, Preziosi E, Madè B, Giuliano G (2006) Natural Arsenic in Groundwater: a case study from central Italy. Mexico- City: International Congress Natural Arsenic in Groundwaters of Latin America. pp 20-24
66. Vivona R, Preziosi E, Made B, Giuliano G (2007) Occurrence of minor toxic elements in volcanic sedimentary aquifers; a case study in central Italy. Hydrogeol J 15:1183-1196
67. Webster TJ (2001) A principal component analysis of the U.S. News & World Report tier rankings of colleges and universities. Econ Educ Rev 20:235-244
68. Webster JG, Nordstrom DK (2003) Geothermal arsenic. In: Welch AH, Stollenwerk KG (eds) Arsenic in ground water, geochemistry and occurrence, Dordrecht, The Netherlands: Kluwer. pp 101-112
69. Welch AH, Stollenwerk KG (2003) Arsenic in ground water: geochemistry and occurrence. Norwell, Massachusetts: Kluwer Academic Publishers. 475 pp`
70. Welch AH, Westjohn DB, Helsel DR, Wanty RB (2000) Arsenic in ground water of the United States. Ground Water 38(4):589-604
71. WHO (2006) Guidelines for drinking water quality. 20 Avenue Appia, 1211 Geneva 27, Switzerland: World Health Organization. pp 488-493

Chapter 6

Cultivation-independent Methods Applied to the Microbial Prospection of Oil and Gas in Soil from a Sedimentary Basin in Brazil

Paula B Miqueletto[1,4], Fernando D Andreote[2], Armando CF Dias[2], Justo C Ferreira[3], Eugênio V dos Santos Neto[3], and Valéria M de Oliveira[1]

[1]Division of Microbial Resources, Research Center for Chemistry, Biology and Agriculture (CPQBA), UNICAMP, CP 6171, CEP 13081-970, Campinas, SP, Brazil

[2]Departament of Soil Sciences, ESALQ, University of São Paulo, CP 09, CEP: 13418-900, Piracicaba, SP, Brazil

[3]PETROBRAS R&D Center, Cidade Universitária, Ilha do Fundão, Rio de Janeiro, RJ, CEP 21949-900, Brazil

[4]Biomedical Sciences Institute (ICB-IV), University of São Paulo, São Paulo, Brazil

ABSTRACT

The upper parts of oil field structures may leak gas which is supposed to be indirectly detected by the soil bacterial populations. Such microorganisms are capable of consuming this gas, supporting the Microbial Prospection of Oil and Gas (MPOG) methodology. The goal of the present work was to characterize microbial communities involved in short-chain alkane metabolism, namely methane, ethane and propane, in samples from a petroliferous (P) soil through clone libraries of the 16S rRNA gene of the Domains Bacteria and Archaea and the catabolic gene coding for the soluble di-iron monooxygenase (SDIMO) enzyme alpha subunit. The microbial community presented high abundance of the bacterial phylum Actinobacteria, which represented 53% of total clones, and the Crenarchaeota group I.1b from the Archaea Domain. The analysis of the catabolic genes revealed the occurrence of seven Operational Protein Families (OPF) and higher richness (Chao = 7; Ace = 7.5) and diversity (Shannon = 1.09) in P soil when compared with a non-petroliferous (Np) soil (Chao = 2; Ace = 0, Shannon = 0.44). Clones related to the ethene monooxygenase (EtnC) and methane monooxygenase (MmoX) coding genes occurred only in P soil, which also presented higher levels of methane and lower levels of ethane and propane, revealed by short-chain hydrocarbon measures. Real-time PCR results suggested that the SDIMO genes occur in very low abundance in the soil samples under study. Further investigations on SDIMOs genes in natural environments are necessary to unravel their still uncharted diversity and to provide reliable tools for the prospection of degrading populations.

INTRODUCTION

Surface geochemical petroleum exploration is defined as the search for migrated surface hydrocarbons and their alteration products, including changes in vegetation and microbial populations [Hitzman et al. 2009; Rashed et al. 2008; Davis and Updegraff 1954; Schumacher 2000]. Microbes play a profound role on the oxidation of migrating hydrocarbons, and are directly responsible for many surface manifestations of petroleum seepage. In this context, the Microbial Prospection for Oil and Gas (MPOG), developed in Germany and used as a stand-alone technique for detecting microseepages since 1961, is based on the knowledge that oil and gas fields emit a continuous stream of light-hydrocarbon gases towards the Earth's surface [Schumacher 2000;Wagner et al. 2002; Tucker and Hitzman 1996].

Specialized microorganisms, such as the hydrocarbon-oxidizing bacteria, depend on light-hydrocarbon gases as their only energy source [Wagner et al. 2002]. In terms of investigation aiming at the microbial hydrocarbon prospection, two groups are relevant: gram-positive bacteria, represented mainly by Actinobacteria from CRNM complex (*Corynebacterium, Rhodococcus, Nocardia* and *Mycobacterium*) that uses short-chain hydrocarbons (C2-C8) as an energy source, and gram-negative bacteria, mainly the genus *Pseudomonas* that possesses the ALK system responsible for alkane degradation [Ginkel et al 1987; Shennan 2006; Kotani et al. 2006].

The Microbial Oil Survey Technique (MOST), developed by Phillips Petroleum Company, is one of the exploration methods and has been available to the petroleum industry since 1985 [Hitzman et al. 2009]. This methodology is based on the isolation of microorganisms on agar plates containing selective growth medium and subsequent counting of colony-forming units. Microbial anomalies have been proven to be reliable indicatives of oil and thermogenic gas occurrences in the subsurface, and the method has been widely used throughout the world. Generally, these methods involve microbial activity analysis in samples taken from depths of 0.2 to one meter [Shennan 2006].

Light hydrocarbon gases that migrate upward from buried reservoirs and become adsorbed to near-surface soils and sediments also represent a useful tool for oil and gas prospection, constituting the Sorbed Soil Gas (SSG) technique [Hitzman et al. 2009]. Areas of microseepage are detected by observing the concentration and composition of light hydrocarbons, chiefly methane through butane, extracted from these soils and sediments. These exploration methods can be used in combination with other data, such as geological and geophysical, to reduce exploration costs and increase success rates [Schumacher 2000]. According to Hitzman (2009), prospections associated with microseepage anomalies are 4-6 times more likely to result in a commercial discovery than prospects with no associated seepage anomaly.

In spite of the well established exploration methods, it is important to emphasize that all of the techniques applied for such goals have been, for many years, based on traditional microbiological methods, encompassing microbial isolation and colony-forming units counting.

Provided the knowledge that only a very small fraction of microorganisms can be recovered in laboratory, this work aimed at a cultivation-independent characterization of the microbial community from a petroliferous basin area explored by PETROBRAS (Brazil), as well as from a randomly selected non-petroliferous area. We also focused on specialized populations of microorganisms, the short chain hydrocarbon-oxidizing bacteria, which are directly involved in light gas metabolism. Soluble di-iron monooxygenases (SDIMOs) were the group of enzymes chosen as a target in order to detect these specialized populations. SDIMOs are represented by multicomponent enzymes that catalyze the initial oxidation of hydrocarbons in phylogenetically and physiologically diverse bacteria [Coleman et al. 2006]. Their structure include a minimum of four proteins; a hydroxylase with two or three subunits, an effector (or coupling) protein and a reductase [Leahy et al. 2003]. The presently biochemically characterized SDIMOs belong to six distinct lineages. The enzymes of subgroups 1 and 2 function predominantly as aromatic monooxygenases and the ones of groups 3-6 as aliphatic monooxygenases. The SDIMOs have numerous applications in bioremediation and biocatalysis [Holmes and Coleman 2008].

MATERIALS AND METHODS

Sampling and Area Background

Soil samples were provided by CENPES/Petrobras (Rio de Janeiro, Brazil). Sampling was performed in September of 2007 on the terrestrial petroliferous field *Jaçanan*, located in Potiguar Basin (Brazil), with the use of appropriate sampling equipment at a 70 cm depth and following strict procedures to avoid contaminations. Samples were then immediately transferred to metal containers and hermetically closed for subsequent gas analysis from the headspace compartment. Sample from this area was named "Petroliferous Soil (P Soil)". Non-petroliferous soil was collected in March of 2007 in the surroundings of Lorena city (Southeast, Brazil) and the same methodology was used to collect and store the samples.

The *Potiguar* sedimentary basin, located in northeast Brazil, is an area of growing exploration interest. Exploration in the basin began onshore in 1956 and nowadays it has approximately 70 oilfields, 58 of which are onshore and account for 85% of the basin's production. The basin currently produces approximately 125,000 barrels of oil equivalent per day, making it the second most important producing area in Brazil after the Campos basin [Lovatini et al. 2010].

Gas Chromatography Analysis

Gas analysis of the soil samples was carried out at PETROBRAS R&D Center (Rio de Janeiro, Brazil). The quantitative analysis of hydrocarbons in the C_1 to C_5 interval were performed by a Hewlett-Packard 5890 series II gas chromatograph equipped with a flame ionization detector (FID) heated at 250°C. A fused silica HP-5-MS capillary column (50 m length, 320 μm ID, 8 μm film thickness) was used. The oven temperature was programmed from 50°C to 200°C with an initial isotherm of 1.75 min and a heating rate of 15°C/min. Helium was used as carrier gas at a constant flow of 8.8 mL/min and quantification was performed using an external calibration with standard gas mixture.

DNA Extraction

DNA extraction from soil samples was carried out using the PowerSoil™ DNA Isolation Kit (Mo Bio Laboratories), according to the manufacturer's instructions. The yield and integrity of the environmental DNA obtained were confirmed through electrophoresis in 1% agarose gel.

16S rRNA Gene Libraries

Two 16S rRNA gene libraries were constructed from the sedimentary soil sample (P soil), one for each domain, Bacteria and Archaea. For the bacterial 16S rRNA gene library construction, the amplification was performed using the primer set 10f and 1100r [Lane 1991]. Fifty μl (50 μl) reaction mixtures contained 50-100 g of total DNA, 2 U of

Taq DNA polymerase (Invitrogen), 1×*Taq* buffer, 1.5 mM $MgCl_2$, 0.2 mM of dNTP mix (GE Healthcare) and 0.4 mM each primer. The PCR amplifications were done using an initial denaturation step of 2 min at 95°C, followed by 30 cycles of 1 min at 94°C, 1 min at 55°C, and 3 min at 72°C, followed by a final extension step at 72°C for 3 min, in an Eppendorf thermal cycler. The archaeal 16S rRNA amplification was performed using the primer set ARCH344f [Casamayor et al. 2002] and ARCH1400r [Kudo et al. 1997] and the following amplification program: an initial denaturation of 94°C for 1.5 min; nine cycles of 30 s at 94°C, 30 s at 67°C (decreasing 0.5°C each cycle) and 1.5 min at 72°C for extension; followed by other 25 cycles of denaturation at 94°C for 30 s, 62°C for 30 s, 72°C for 1.5 min and a final extension step of 10 min at 72°C.

SDIMO Gene Libraries

A nested PCR strategy and degenerate primers were employed for the amplification of the gene coding for the alpha subunit of soluble di-iron monooxygenase enzymes from the environmental samples and subsequent library construction. In this case, one gene library was assembled for each soil sample, the petroliferous (P soil) and non-petroliferous soil (Np soil). In the first PCR reaction total genomic DNA from soil sample was used as template and the primer set NVC58 and NVC65 [Coleman et al. 2006] was used, yielding a fragment with an expected of 1,100 bp. The PCR product from the first reaction was then employed as template for the second PCR amplification using the primer set NVC66 and NVC57 [Coleman et al. 2006], which yielded a smaller fragment with an expected size of approximately 410 bp. The primers were designed to target conserved regions of the gene coding for the alpha subunit of subgroups 3, 4, 5 and 6, which are involved in the degradation of alkenes and propane. PCR conditions were as described inColeman et al. (2006).

Cloning and Sequencing of PCR Products

PCR replicates of each target gene were pooled, purified using GFX PCR DNA and Gel Band Purification kit (GE Healthcare), according to the manufacturer's protocol, and concentrated in a speed vacuum concentrator 5301 Eppendorf, A-2-VC rotor. The purified PCR product (200 g) was cloned into a pGEM-T Easy Vector (Promega), according to the manufacturer›s instructions, and transformed into *E. coli* JM109 competent cells. Approximately 200 positive clones were selected from each library for subsequent sequencing. The 16S rDNA and SDIMO inserts were amplified from plasmid DNA of selected clones using the universal the M13 forward (5′-CGCCAG GGT TTT CCC AGT CAC GAC-3′) and reverse (5′-TTT CAC ACA GGA AAC AGC TAT GAC-3′) primers. PCR was performed in a 50-μl reaction volume, containing 1-2 μl of an overnight clone culture, 0.4 μM of each primer, 0.2 mM dNTP mix, 2 U *Taq* DNA polymerase (Invitrogen), 1× *Taq* buffer, and 1.5 mM $MgCl_2$. The amplification program consisted of an initial denaturation step at 94°C for 3 min, followed by 30 cycles of 94°C/20 s, 60°C/20 s, and 72°C/90 s. PCR products were purified as previously described for automated sequencing in the MegaBace DNA Analysis System 500 (GE Healthcare). The sequencing was carried out using 10f and 1100r primers for bacterial rRNA 16S; 344f and 1400r for Archaeal rRNA 16S, and NVC57 and NVC66 for the catabolic genes.

Phylogenetic Analysis

Partial 16S rRNA sequences obtained from clones were assembled in a contig using the phred/Phrap/CONSED program [Ewing et al. 1998]. Identification was achieved by comparing the contiguous 16S rRNA sequences obtained with 16S rRNA sequence data from reference type strains, as well as environmental clones available in the public databases GenBank (http://www.ncbi.nlm.nih.gov) and RDP (Ribosomal Database Project - Release 10;http://rdp.cme.msu.edu) using the BLASTn and Classifier routines, respectively. The sequences were aligned using the CLUSTAL X program [Thompson et al. 1997] and analyzed with MEGA software v.4 [Kumar et al. 2001]. Evolutionary distances were derived from sequence-pair dissimilarities calculated as implemented in MEGA, using Kimura's DNA substitution model [Kimura 1980]. For the identification of catabolic gene sequences BLASTx routine was used and nucleotide sequences coding for the alpha subunit SDIMO gene were translated *in silico*. The phylogenetic reconstruction was done using the neighbor-joining (NJ) algorithm, with bootstrap values calculated from 1,000 replicate runs, using the routines included in the MEGA4 software.

Statistical Analysis

Diversity index calculations (-diversity measures) were performed individually for all gene libraries with the programs MOTHUR version 1.13.0 and DOTUR v. 1.3 [Schloss & Handelsman 2005]. DOTUR was used to assign bacterial and archaeal 16S rRNA sequences to OTUs and to calculate rarefaction curves using 4 distance levels among sequences (3, 5, 10 and 20%), and to calculate, for each community, the Shannon (H›) diversity index, Simpson index [Magurran 2004], and the nonparametric richness estimators ACE (ctimator) [Chao 1993] and Chao1 [Chao 1984]. Chao1 richness estimates were based on singletons and doubletons as described by Chao (1984), and ACE was based on the distribution of abundant (>10) and rare (≤10) species. MOTHUR was used for the analysis of SDIMO gene sequences, supporting the definition of Operational Protein Families (OPF) at a distance level of 83%.

Quantitative PCR

Real time PCR was carried out in an Applied Biosystems thermo cycler (Step One Real Time PCR System). The reaction mixture was prepared using the Superscript® III Platinum® SYBR®Green One-Step qPCR Kit w/ ROX (Invitrogen), according to the manufacturer›s recommendations. The primer set BacqPCR - P1 (5› -CCT ACG GGA GGC AGC AG - 3›) and BacqPCR - P2 (5› - ATT ACC GCG GCT GCT GG - 3›) was employed for the bacterial 16S rDNA amplification, and the primer set NVC57 and NVC66 [Coleman et al. 2006] was used for the catabolic gene amplification. The real-time PCR program for the target DNA amplification was performed as follows: 95°C for 2 min followed by 30 cycles of 1 min at 94°C, 1 min at 55°C and 3 min at 72°C for extension. Bio-Rad iQ5 Optical System software (version 1.0, 2005) was used to analyze the amplification data.

Standard Curve of Real-Time PCR

PCR products containing SDIMO and 16S rRNA gene fragments were used separately for each standard curve construction. Tenfold serial dilution (from 10^{-1} to 10^{-7}) was used in triplicate as PCR templates in order to produce curves to enable the quantification of total populations of bacteria as well as of short-chain hydrocarbon degrading microorganisms.

Nucleotide Sequence Accession Numbers

The 16S rRNA and SDIMO gene partial sequences determined in this study for the environmental clones were deposited at the Genbank database under the accession numbers JF681794 toJF681937 and JF694327 to JF694383 for the 16S rRNA from bacteria and archaea respectively and the numbers JN116289 to JN116426 for the SDIMO gene.

RESULTS

Chromatographic Analysis

Chromatographic data revealed relatively high concentration of methane in the soil sample from the petroliferous field (P soil) in comparison with ethane and propane contents (Table 1). For a comparison purpose, one soil sample from a different area, named "Non-petroliferous soil (Np Soil)", was submitted to gas chromatography analysis as well. This sample was collected in a region outside the sedimentary basin and represents a type of soil that would, in thesis, harbor lower levels of short-chain hydrocarbons, especially those related to the presence of petroleum at the subsurface of the studied area. These qualitative data function as an information input for our analysis, since it gives an idea of light gas levels that one can expect to find in samples with different characteristics such as these. The present analysis showed that, surprisingly, the sample representing the non-petroliferous area contained higher concentrations of ethane and propane, components of natural gas from petroleum, in comparison with those found in the P soil (Table 1).

Table 1: Measures of short-chain hydrocarbons contents taken from the head-space of Np Soil and P Soil containers (ppm)

	P Soil	Np Soil
Methane	1152.14	140.36
Ethane	1.16	5.36
Propane	0.39	3.43

Miqueletto et al. AMB Express 2011 1:35 doi: 10.1186/2191-0855-1-35

Composition of the Soil Communities

The composition of bacterial and archaeal communities was determined by analysis of 16S rRNA gene clone libraries. A total of 160 partial sequences were obtained from the 16S rRNA bacterial library and 60 from the archaeal library, each consisting of 850 nucleotides in average. These 16S rRNA sequences were compared with sequences from reference and type strains, as well as environmental clones, available at the GenBank and RDP II databases. Ten major bacterial taxa were identified (Figures 1A and 1B), being the phylum Actinobacteria the most abundant one, accounting for 52.2% of all bacterial clones distributed among the orders Actinomycetales (25%), Rubrobacterales (9.5%) and Solirubrobacterales (0.7%). Clones classified as uncultured Actinobacterium accounted for 17% of the bacterial library (Figure 1A).The second most abundant was the phylum Proteobacteria, which represented 12.5% of the bacterial library (2.5% affiliated to Alphaprotebacteria, 2% to Betaproteobacteria, 3.7% to Gammaproteobacteria and 4.3% to Deltaproteobacteria); followed by Acidobacteria representing 11% of total clones. The remaining sequences were significantly less numerous, being related to the phyla Gemmatimonadetes (5.8%), Bacteroidetes (2.5%), Chloroflexi (2.5%), Cyanobacteria (2.5%), Firmicutes (2%) and Verrucomicrobia (0.6%). Some sequences (6.4%) could not be affiliated to any known taxa (Figures 1A and 1B).

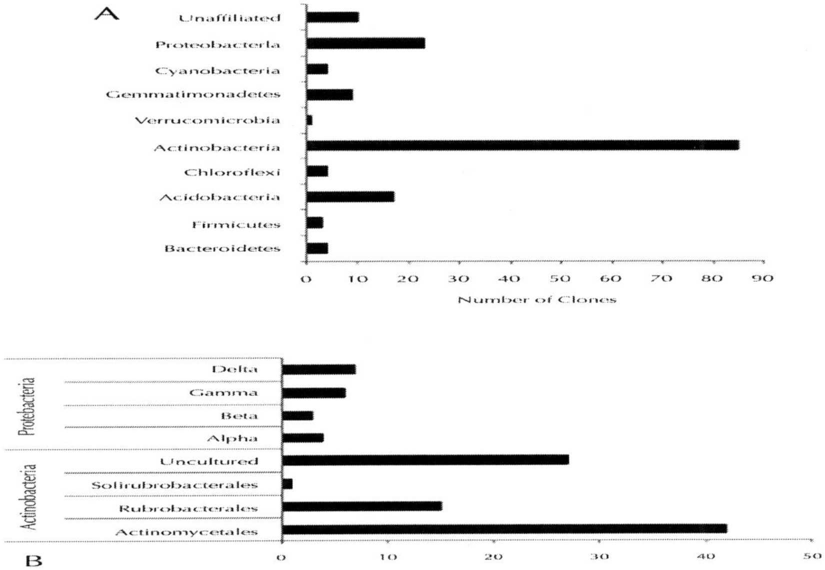

Figure 1: Proportion of 16S rRNA gene clones from P Soil among bacterial phyla (A); among classes within the phylum Proteobacteria and among orders within the phylum Actinobacteria (B). Sequence classification was based on the RDP classifier and BLASTn results.

Phylogenetic analysis allowed identification of many bacterial members of the soil sample at the species level (Figures 2 and 3). The majority of Actinobacteria-related clones occurring in the 16S rDNA library from the petroliferous soil were clearly related to the order Actinomycetales (Figure2), which includes species from the CNMR complex, such as *Mycobacterium* sp. TY-6. Several clones were closely related to species like *Sporichthya polymorpha, Saccharopolyspora antimicrobica, Saccharopolyspora gloriosa, Nocardioides alkalitolerans* and *Marmoricola aequoreus*. Rubrobacterales was the second more abundant order within the phylum Actinobacteria, represented in the tree by three clones closely related to *Rubrobacter radiotolerans*. The orders Solirubrobacterales and Acidimicrobiales were also identified in the sample and were represented by the species *Solirubrobacter soli* and *Iamia majanohamensis*, respectively.

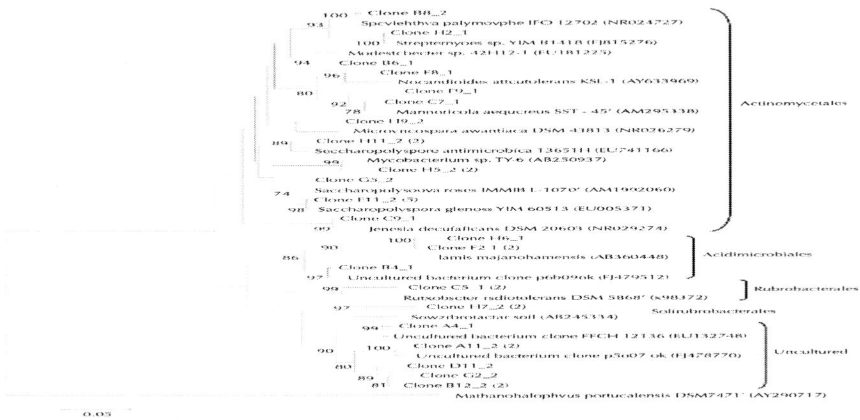

Figure 2: Phylogenetic analysis based on partial 16S rRNA sequences from the P Soil clone library representing members of phylum Actinobacteria and related species. Evolutionary distances were based on the Kimura 2p model and tree reconstruction on the neighbor-joining method. Bootstrap values (1,000 replicate runs, shown as %) greater than 70% are listed. GenBank accession numbers are listed after species names. *Methanohalophilus portucalensis* was used as the outgroup.

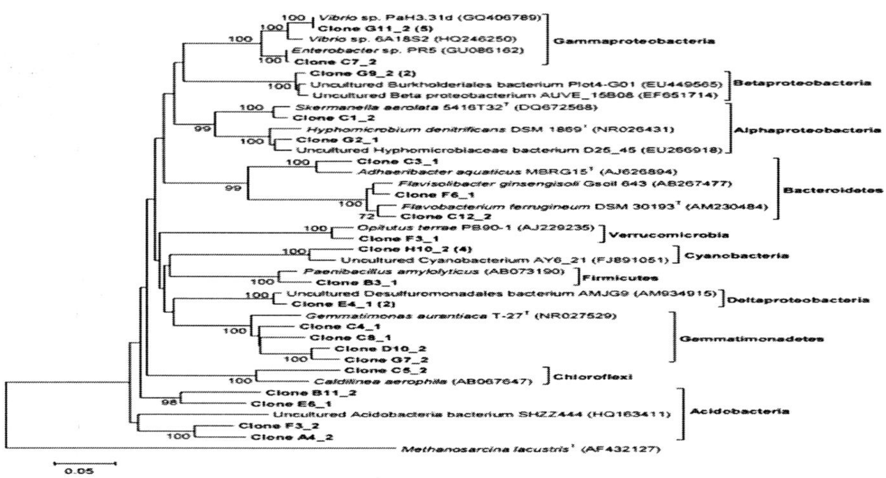

Figure 3: Phylogenetic analysis based on partial 16S rRNA sequences from P Soil clone library representing members of major phyla in the Bacteria Domain, except Actinobacteria, and related species. Evolutionary distances

were based on the Kimura 2p model and tree reconstruction on the neighbor-joining method. Bootstrap values (1,000 replicate runs, shown as %) greater than 70% are listed. GenBank accession numbers are listed after species names. *Methanosarcina lacustris* was used as the outgroup.

Clones affiliated to Proteobacteria were closely related to representatives of each class of the group, supported by high bootstrap values (100%), as shown in Figure 3. Deltaproteobacteria was represented by the order Desulfuromonadales. Several 16S rDNA sequences were affiliated to Gammaproteobacteria, identified as *Vibrio* sp., while sequences affiliated to Betaproteobacteria were related mainly to the order Burkholderiales. Some clones affiliated to the class Alphaproteobacteria were identified at the species level, such as *Hyphomicrobium denitrificans* and*Skermanella aerolata*.

Other bacterial species identified in this study included *Paenibacillus amylolyticus*, affiliated to Firmicutes, *Caldilinea aerophila*, representing the Chloroflexi phylum, and *Opitutus terrae*, belonging to the phylum Verrucomicrobia. The phylum Bacteroidetes was represented by clones related to the species *Flavisolibacter ginsengisoli, Flavobacterium ferrugineum*, and *Adhaeribacter aquaticus*.

The phylum Gemmatimonadetes was represented in the P Soil 16S rDNA library by four clones identified as *Gemmatimonas aurantiacus*.

The sequencing of the 16S rRNA archaeal library revealed that 56% of the community in P soil was represented by the Crenarchaeota phylum, with clones being identified as uncultured Crenarchaeota (Figure 4). Approximately 42% of total clones were included in the category*Candidatus nitrososphaera gargensis*, currently classified as a member of the phylum Thaumarchaeota. Only 2% of the sequences were afilliated to a species from the Euryarchaeota group, *Methanosarcina acetivorans*, a methanogenic organism.

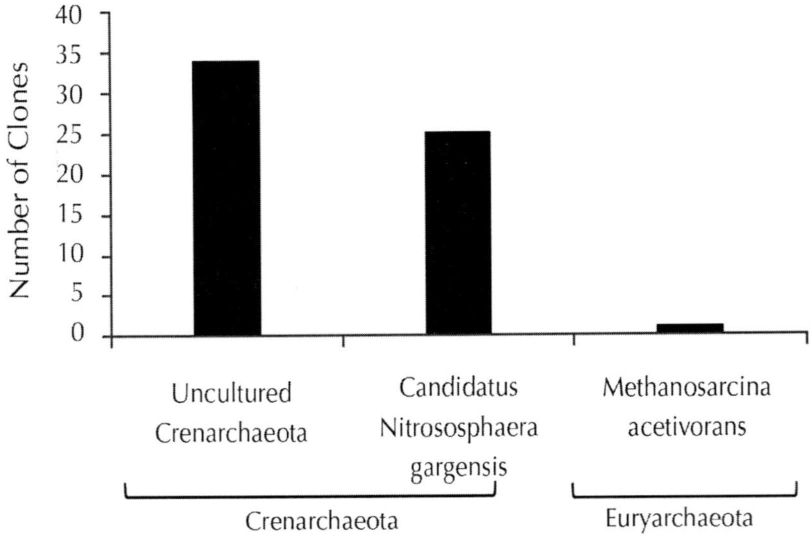

Figure 4: Proportion of 16S rRNA gene clones from P Soil among archaeal phyla. Sequence classification was based on the RDP classifier and BLASTn results.

SDIMO Gene Diversity in Soil Samples

Sixty eight SDIMO alpha subunit gene sequences derived from P Soil and seventy from Np Soil were sequenced and analyzed. For a preliminary analysis, these sequences were aligned in order to detect changes in the nucleotide sequence and a phylogenetic tree was constructed (data not shown). It is worth to mention that the great majority of the sequences from the soil samples under study were clustered separately from reference sequences, showing to be more closely related to each other. The cut-off value of 83% was selected to define distinct operational protein families (OPF) based on the analysis of the phylogenetic trees of total clones in combination with Mothur analysis. Seven OPFs were defined for the P Soil, whereas only two OPFs were defined for the non-petroliferous soil. Representative sequences of distinct OPFs were submitted to analysis using the BLASTx routine. Sequences considered as belonging to the same OPF matched the same sequences from the GenBank database, with slight differences in identity values and e-values. It was interesting to notice that, in spite of housing such catabolic genes, the Np soil sample presented lower OPF diversity in comparison to P soil (Figure 5).

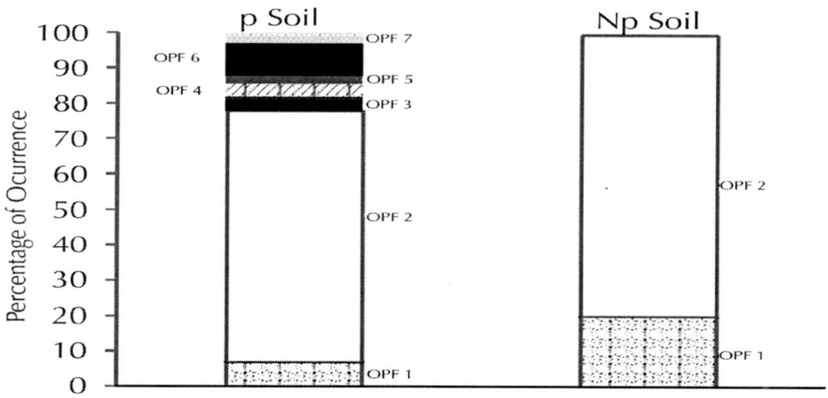

Figure 5: Soluble monooxygenase alpha subunit gene frequency and OPF identification in soil samples from petroliferous (P) and non-petroliferous (NP) areas.

Most of the SDIMO sequences were distributed among OPFs 1 and 2 and were related to yet uncultured bacteria, being OPF 2 the most frequent one which comprised 48 sequences in P Soil and 56 in Np Soil (Table 2). All the OPFs matched SDIMO gene sequences with high similarity values (66% to 93.4%), especially with gene sequences from uncultured bacteria. Lower identity values were observed relating the query sequences to propane monooxygenase gene from the genus *Mycobacterium*. Some clones from P Soil sample contained genes identified as coding for methane monooxygenase (MmoX) from *Nocardioides* sp. and ethene monooxygenase (EtnC) from*Mycobacterium chubuense*.

Table 2: Blastx analyses of *SDIMO* clone sequences representing each of the OPF determined in the present study and the highest identity matches from the Genbank database

OPF	Number of clone sequences	Gene	Best match	Average of Identity (nt)
1	5 (P) - 14 (Np)	SDIMO	Uncultured bacterium ABB70434	81.4%
2	48 (P) - 56 (Np)	SDIMO	Uncultured bacterium ABB70441	93.2%
3	3 (P)	SDIMO	Uncultured bacterium ABB70469	93.4%
4	3 (P)	EtnC	Mycobacterium chubuense NBB4	83.7%
5	1 (P)	SDIMO	Mycobacterium sp. JS623	66.0%
6	6 (P)	SDIMO	Uncultured bacterium ABB70430	77.4%
7	2 (P)	MmoX	Nocardioides sp. JS614	88.5%

MmoX represents Methane Monoxygenase, EtnC represents Ethene Monoxygenase, and SDIMOs represents soluble di-iron monoxygenase enzymes that have not been assigned to a known subgroup. Identity average shows the percentage of identity at the nucleotide level (nt) compared with sequences from Genbank database. (P) Represents sequences from P soil and (Np) from Np soil.

Miqueletto et al. AMB Express 2011 1:35 doi: 10.1186/2191-0855-1-35

Catabolic gene sequences were translated *in silico* and representative sequences of each defined OPF were selected and together with SDIMO alpha subunit amino acid sequences from reference microorganisms were used for the tree reconstruction. Some of these sequences showed high sequence similarity levels with sequences from yet uncultured microorganisms available in the database and were recovered in tight clusters supported by high bootstrap values (Figure 6). Actually, only a few SDIMO sequences grouped with sequences from isolated microorganisms, distributed among OPFs 4, 5 and 7. These sequences showed lower similarity values, being only distantly related to known taxa such as *Nocardioides* and *Mycobacterium*. Since the reconstruction of the phylogenetic tree was based on amino acid sequences, sequences from OPFs 2 and 6, which had presented minor differences in nucleotide bases in the ClustalX alignment, were recovered in a single cluster supported by 100% bootstrap value (Figure 6). It is known that subtle changes among nucleotide sequences coding for SDIMO enzymes are considered essential because they can generate distinct biochemical functions [Cooley et al. 2009]. Therefore, based on the analyses performed using ClustalX and Mothur, the authors chose not to neglect these differences when defining the OPFs.

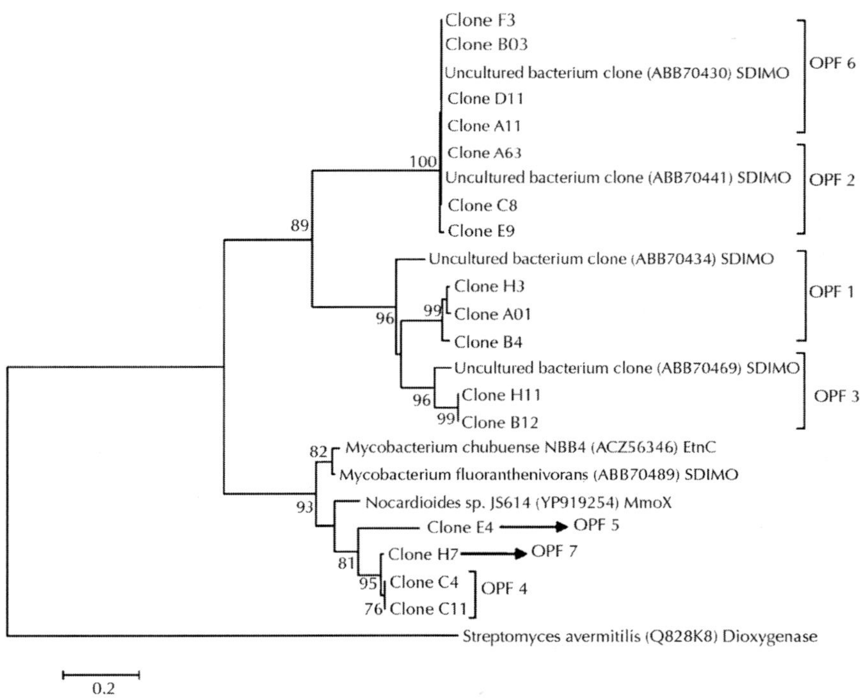

Figure 6: Phylogenetic tree based on the alignment of amino acid sequences of the alpha subunit of soluble di-iron monooxygenases from clones from P soil sample and related sequences from the database. Access numbers of *SDIMO* genes from reference strains are in brackets.

Richness and Diversity Analyses

DOTUR analyses of the 16S rDNA bacterial sequences from P soil sample allowed the definition of 105 OTUs from 145 good quality sequences, generating 83 singletons and 14 doubletons. The highest frequency OTU grouped only six sequences. Rarefaction analysis yielded an asymptotic curve at the distance level of 0.2, meaning that if more sequences were analyzed, new phyla would not have been detected (Figure 7A). However, when the distance level of species is considered (D = 0.03), it becomes clear that an increase of the sampling effort would allow the identification of new species in the data set (Figure 7A).

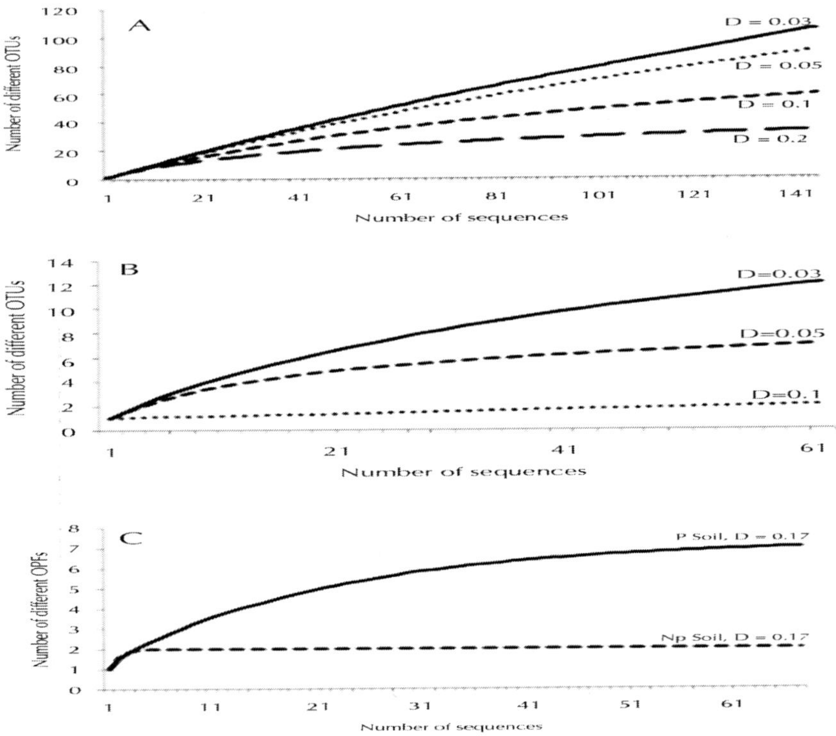

Figure 7: Rarefaction analyses of 16S rRNA gene sequences from bacteria (A) and archaea (B), and of SDIMO gene sequences (C) obtained from P and Np Soil samples.

Diversity index and richness estimator values generated by DOTUR corroborated the great bacterial species diversity and richness that would be expected in a soil community (Table 3). In spite of the Actinobacteria abundance, the high value of Shannon index at the distance level 0.03 indicates that a great OTU diversity is also present within this phylum.

Table 3: Richness and diversity estimates for bacterial and archaeal 16S rRNA and SDIMO gene clone libraries from both P and Np Soil samples (sequence classification based on the cutoff value determined by DOTUR)

Gene (na)	Richnessb	Distancec	Chaod	Acee	Shannonf
Bacterial 16S rDNA library (145)	105	0.03	331.8	391.4	4.49
Archaeal 16S rDNA library (53)	12	0.03	17	18.3	2.05
SDIMO Gene (68 - P Soil)	7	0.17	7	7.3	1.09
SDIMO Gene (70 - Np Soil)	2	0.17	2	0	0.44

[a]n, Number of gene sequences analyzed.

[b]97% identity was estimated as the species-level distance (D = 0.03) for Bacteria and Archaea, and 83% identity for SDIMO genes.

[c]Richness is based on observed unique OTUs.

[d]Nonparametric statistical prediction of total richness of different OTUs and OPFs based on distribution of abundant (10) and rare (10) OTUs.

[e]Nonparametric statistical predictions of total richness of OTUs and OPFs based on distribution of singletons and doubletons.

[f]Shannon diversity index. A higher number represents more diversity.

Miqueletto et al. AMB Express 2011 1:35 doi: 10.1186/2191-0855-1-35

The analysis of the archaeal 16S rRNA gene library, allowed the definition of 12 OTUs out of 53 good quality sequences. Rarefaction curves at the estimated phylum and order levels also reached clear saturation, suggesting that the sampling effort was sufficient to reveal all archaeal phyla and orders present in this sample (Figure 7B). Five singletons were generated in the analysis and the most frequent OTU presented 16 sequences. These results were corroborated by the

Shannon diversity index and ACE and Chao richness estimator values, which were significantly lower than those found for the bacterial community (Table 3).

A total of 68 good quality sequences derived from the catabolic gene libraries from P Soil were analyzed and seven distinct protein groups were obtained based on the distance level previously defined for OPFs, which were later confirmed by Blast score ratio analysis performed in BlastP database. For the Np soil sample, 70 sequences were analyzed and only two protein groups could be defined and confirmed. Rarefaction analysis performed for the SDIMO gene library from both samples yielded plateau-shaped curves, indicating that the saturation of the gene diversity in the environment was complete and the sampling effort was satisfactory, in particular for the Np soil sample (Figure 7C).

Real Time Quantitative PCR

With the use of the real-time PCR technique it was possible to successfully construct the bacterial 16S rRNA gene standard curve, achieving satisfactory indices for bacterial quantification (R2 = 0.99, efficiency = 94%). The DNA sample representing the target to be quantified reached the interval of the generated curve among the points from the dilution series. The log copy number calculated was 6.901/g and 7.427/g soil for P Soil and Np Soil, respectively (Figure 8).

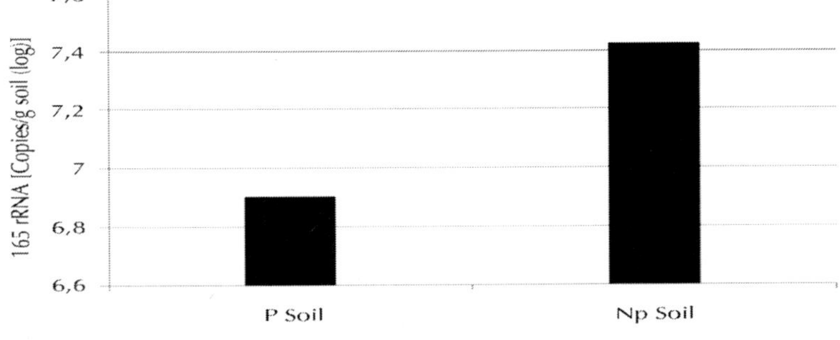

Figure 8: Quantification of bacterial 16S rRNA gene in both soil samples analyzed. P soil indicates the soil at Potiguar Basin, and Np soil indicates the non-petroliferous soil.

The application of this technique aimed at the quantification of the total bacterial population, through the 16S rRNA gene copy number, and the quantification of specialized populations of microorganisms that consume short-chain hydrocarbons, through the SDIMO gene quantification. Such analysis would have generated the proportion of potentially degrading bacterial population in relation to the total community. However, quantification of SDIMO genes yielded a very low copy number, which was probably under the detection level of the analytic procedure.

DISCUSSION

Prokaryotic community diversity and abundance were investigated in a soil sample originated from a sedimentary basin by means of 16S rRNA and catabolic gene libraries in combination with real-time PCR. Results revealed the outstanding high abundance of the phylum Actinobacteria in the bacterial community, especially the order Actinomycetales, an important group inhabiting the soil environment that encompasses the genera from the CRNM complex, known for its hydrocarbon degradation skills [Shennan 2006].

In general, most of the groups identified in the samples are of common occurrence in the environment, especially in soils and sediments. In the Actinobacteria phylum, known for its ability of surviving under adverse conditions, the genus *Saccharopolyspora* had a considerable contribution. Clones were affiliated to the species *Saccharopolyspora rosea,* which was first isolated from a patient with bronchial carcinoma [Yassin 2009], and other species that generally occur in the environment, such as saline lakes, soil, marine sediments and even sponge tissue:*Saccharopolyspora gloriosa, Saccharopolyspora cebuensis, Saccharopolyspora antimicrobica, Saccharopolyspora salina* and *Saccharopolyspora halophila* [Pimentel-Elardo et al. 2008; Yuan et al. 2008; Suthindhiran & Kannabiran 2009; Tang et al. 2009; Qin et al. 2010]. Some other species belonging to the same order, Actinomycetales, included *Actinopolyspora xinjiangensis,* an extreme halophile isolated from salt-lake sediments [Guan et al. 2010], *Nocardioides aestuarii,* isolated from tidal flat sediment [Yi and Chun 2004] and *Jonesia denitrificans,* a pathogenic organism that had its genome sequencing completed in 2009 [Pukall et al. 2009].

Some genera and species more specifically related to the metabolism of hydrocarbons were also recovered, such as *Mycobacterium pallens*, a polycyclic aromatic hydrocarbon-degrader isolated from Hawaiian soils [Hennessee et al. 2009], *Mycobacterium* sp. TY-6, known for its capacity of degrading propane [Kotani et al. 2006] and *Nocardioides* sp., an alkene-degrading bacteria [Vomberg and Klinner 2000]. In this sense, there was also a contribution of organisms from the phylum Proteobacteria, such as uncultured bacteria from the family Burkholderiaceae (Betaproteobacteria), which comprises oil degrading lineages and cultures reported to degrade both aliphatic and aromatic hydrocarbons [Chakraborty et al. 2010], and an *Enterobacter* sp. strain (Gammaproteobacteria), which is known for its ability of degrading methyl tert-butyl ether (MTBE), a gasoline constituent [Kao et al. 2010]. The order Desulfuromonadales (Deltaproteobacteria), constituted mainly by anaerobic bacteria [Thomas et al. 2008], was also detected. This group encompasses mainly sulfate-reducing bacteria [Borin et al. 2009] and several species possessing this metabolism may present in addition the ability of degrading short-chain hydrocarbons [Kniemeyer et al. 2007].

The group *Vibrio* was present in the sample and represented mainly by the species *Vibrio chagasii*and *Vibrio splendidus*. Organisms belonging to this genus often move from mutualistic to pathogenic associations with aquatic animals [Lacoste et al. 2001; Gomez-Leon et al. 2005; Venier et al. 2011]. In a recent study, Chien and co-workers (2007) identified *Vibrio* strains from mangrove sediment able to accumulate polyhydroxyalkanoates (PHAs). These compounds are polyesters accumulated by various bacteria under unbalanced growth conditions which can be useful for the development of non-petroleum based biodegradable plastics [Chien et al. 2007].

Clones related to the genus *Paenibacillus* (phylum Firmicutes) were also found in the 16S rDNA gene library from P soil sample. Bacteria from this phylum are ubiquitous and have the ability to adapt to different types of environment and generally present high abundance in samples from anaerobic digesters, anaerobic sludge, landfill and fecal samples [Krause et al. 2008; Krober et al. 2009], suggesting the ability to consume complex organic compounds, such as polysaccharides, being essential for organic matter breakdown process and methane production.

Representants of Acidobacteria were identified in P soil sample only as uncultured organisms. The majority of the bacteria from this group are uncultured and only known by their 16S rDNA sequences [Quaiser et al. 2003]. Based on their phylogenetic diversity and ecological distribution, Acidobacteria are expected to represent a metabolically and genetically diverse group comparable to other well known phyla and are currently found in soils, freshwater habitats, hot springs, wastewaters and others [Quaiser et al. 2003].

Gemmatimonadetes-related clones identified in the P soil library presented high sequence similarity to an organism isolated from a community of a semiarid lead-zinc mine tailing site [Mendez et al. 2008]. Representatives of this group are generally isolated from alkaline environments with moderate temperatures and aerobic conditions, and also present low substrate versatility [Hartman 2009].

In spite of the great bacterial diversity frequently found in soils, and also supported in the present work, it is possible to recognize a certain pattern regarding the contribution of each phylum, being the great majority of clones usually affiliated to nine major bacterial phyla and generating, to a certain extent, stability in the community structure. According to Janssen (2006), dominant phyla usually correspond to approximately 92% of the libraries, being Proteobacteria and Acidobacteria the most abundant ones, while Actinobacteria would only represent about 13% of total clones. Phylogenetic analysis of bacterial community from Potiguar petroliferous basin soil sample showed a different distribution in comparison to the general pattern. Actinobacteria had a contribution of more than half of the library, while Proteobacteria and Acidobacteria represented only 12% and 11%, respectively. Several theories have been suggested to explain the significant occurrence of Actinobacteria in environmental samples, being resource limitation the most accepted one, such as the ability to adapt to recalcitrant substrate [Fierer et al. 2007].

The calculation of the Shannon index reinforced the observation that the P soil sample harbors a highly diverse bacterial community. Shannon values found in the literature for bacterial communities ranged from 2.96 to 3.83 in libraries from water column [Dimitriu et al. 2008]; 2.7 to 3.5 in coral samples [Bourne et al. 2008], 3.6 to 4.0 for refinery WWTP sludge [Silva et al. 2010] and a diversity index of 2.9 was obtained for a petroliferous reservoir sample [Li et al. 2007].

According to ecological theories more diverse communities may offer a more significant contribution to the ecosystem functions, keeping in mind that ecological studies on soil microbial communities have revealed that this type of habitat generally presents extreme diversity, with species richness estimates ranging from values of 2000 to 52000 [Youssef & Elshahed. 2009].

A great number of clones from both libraries showed high similarity levels to gene sequences originated from uncultured bacteria and archaea of different environmental samples, such as deep-lake sediment, springs, tropical forest plants, pasture, agricultural soil, marine biofilm, rhizosphere, urban aerosol and bioreactors, suggesting the cosmopolitan distribution of these organisms in the environment.

Phylogenetic analysis of archaeal 16S rRNA gene sequences showed a high abundance of organisms identified as *Candidatus nitrososphaera gargensis*. This group was considered the first member of Crenarchaeota I.1b group in which ammonium oxidation could be observed at the cellular level [Hatzenpichler et al. 2007]. Due to their phylogenetic and metabolic characteristics these organisms are now classified as members of the phylum Thaumarchaeota, which encompasses mesophilic archaea previously affiliated to Crenarchaeota and with a major role in geochemical cycles [Brochier-Armanet et al. 2008; Spang et al. 2010]. *Nitrososphaera gargensis* is a representative of an evolutionary lineage within the group Thaumarchaeota that occurs predominantly in terrestrial environments.

Nonetheless, sequence identification through sequence databases offered significant limitations in the present study and identification at species level turned out to be a very difficult task for a large quantity of sequences. Analysis based on 16S rRNA gene sequencing can be very useful in surveying dominant groups in samples under study, however it can be inefficient in detecting and identifying species related to light gas metabolism, already known for its low abundance in natural environments [Coleman et al. 2006]. Therefore, the analysis of microbial community composition based on 16S rRNA genes libraries of relative small size, in spite of representing a fundamental method for a general characterization of microbial communities, seems to be inefficient for a more specific application such as the microbial prospection of oil and gas. The detection of organisms that correspond to the most abundant OTUs requires minimal sampling, whereas the recovery of sequences from minor components demands surveys that are many orders of magnitude larger than those reported in the literature. These low-abundance organisms that represent rare OTUs constitute the so called "rare biosphere", which is still largely unexplored. According to Sogin (2006), the rare biosphere may serve as a potentially inexhaustible reservoir of genomic innovation and its members usually have a sparse distribution in nature. Recent high-sensitivity surveys of diverse environmental samples have revealed that the vast majority of the microbial diversity in a specific habitat is comprised by taxa that are present at very low abundances and the most abundant organisms represent only a fraction of the total diversity [Brazelton 2010]. In a recent work, Galand and co-workers (2009) demonstrated patterns of biogeography for the rare biosphere and concluded that its diversity is most likely subjected to ecological processes such as selection, speciation, and extinction.

In this context, the use of 16S rRNA gene libraries for the investigation of microbial populations responsible for short-chain alkane degradation, mainly ethane and propane, may be time consuming and offer low resolution in terms of species and genes as key players in the degradation processes. The analysis of catabolic gene libraries may offer a more appropriate technique to be applied to MPOG, considering that it is directly focused on light gas metabolism. The present study revealed that most of the SDIMO genes recovered from soil samples correspond to potentially new genes, which are not related to those found in known bacterial species, a fact that made it difficult to infer which microbial group carried the short-chain alkane degradation genes. SDIMOs subgroups 3, 4, 5 and 6 could be identified in the soil samples under study. The six known subgroups are phylogenetically distinct and present significant evolutionary distance among each other. These large evolutionary gaps suggest that known SDIMOs are not representative of the naturally occurring diversity, what makes the group a worthy target for bioprospecting [Holmes & Coleman 2008]. According to OPFs definition and Mothur analysis it could be inferred that there are fewer lineages of microorganisms involved directly in the metabolism of short-chain hydrocarbons in a non-petroliferous soil.

Hydrocarbon levels measured in P and Np soil samples through gas chromatography did not detect a geochemical anomaly on soil from petroliferous basin as would be expected. Some characteristics from soils have been proved to affect this type of analysis. According to Horvitz (1985) the acidic nature of some soils may prevent the hydrocarbons from being adsorbed, therefore making the real levels that occur in nature different from those measured in laboratory. Another problem is the presence of large quantities of biogenic methane in most near-surface soils, which can confuse the interpretation of gas levels related to the leakage from buried reservoirs.

One of the first works on MPOG based on molecular biological methods, carried out by Zhang and co-workers (2010), focused on soil methanotrophic populations as a bioindicator for the presence of oil and gas in a field in China. In spite of the successful quantification of such organisms by real-time PCR, the common presence of biogenic methane in soil represents an obstacle to link the occurrence of methanotrophic populations to the presence of oil and gas reservoirs. Since methane is a common product of microbial action on organic matter, the presence of methanotrophic organisms in a given sample is less probable of representing an indicative of a reservoir than the occurrence of more specific bacterial populations, such as ethane, propane and butane degraders [Shennan 2006]. In addition, short-chain hydrocarbon degrading bacteria from the CNMR complex, which carry the SDIMO genes, posses several physiological features that make these organisms a more appropriate bioindicator in microbial prospection methods. Slow development in low-nutrient conditions, a peculiar cell wall highly resistant to desiccation, capacity of surviving in starving conditions for long periods and an adjustable endogenous metabolism are some skills that can explain the surviving success of this group.

In conclusion, the results gathered in this work suggest that cultivation-independent microbial molecular techniques in combination with the light hydrocarbon degrading bacterial populations as a target will improve the accuracy rate of MPOG. Further studies on the occurrence and diversity of SDIMO genes in soil, as well as the improvement of primer sets to be applied in real-time PCR, are necessary in order to overcome the obstacle of the low abundance of catabolic genes in natural environments and enable their quantification from complex genetic backgrounds.

ACKNOWLEDGEMENTS

We thank PETROBRAS for the financial support, donation of samples and authorization to publish this work, and to the Chemist Rosane A. Fontes from the GC-lab (PETROBRAS R&D Center/Geochemistry). P. B. M. was supported by grants from CAPES (Coordenação de Aperfeiçoamento de Pessoal de Nível Superior, Brazil).

REFERENCES

1. Borin S, Brusetti L, Daffonchio D, Delaney E, Baldi F (2009) Biodiversity of prokaryotic communities in sediments of different sub-basins of the Venice lagoon. Research in Microbiology 160:307-314
2. Bourne D, Lida Y, Uthicke S, Smith-Keune C (2008) Changes in coral-associated microbial communities during a bleaching event. The ISME Journal 2:350-363
3. Brazelton WJ, Brazelton WJ, Ludwig KA, Sogin ML, Andreishcheva EN, Kelley DS, Shen C, Edwards RL, Baross JA (2010) Archaea and bacteria with surprising microdiversity show shifts in dominance over 1,000-year time scales in hydrothermal chimneys. Proc Natl Acad Sci USA 107:1612-1617
4. Brochier-Armanet C, Boussau B, Gribaldo S, Forterre P (2008) Mesophilic crenarchaeota: proposal for a third archaeal phylum, the Thaumarchaeota. Nature Reviews Microbiology 6:245-252
5. Casamayor EO, Massana R, Benlloch S, Overeas L, Diez B, Goddard V, Gasol JM, Joint I, Rodríguez-Valera F, Pedrós-Alió C (2002) Changes in archaeal, bacterial, and eukaryal assemblages along a salinity gradient by comparison of genetic fingerprinting methods in a multipond solar saltern scaling plots. Appl Environ Microbiol 4:338-348
6. Chakraborty S, Mukherji S, Mukherji S (2010) Surface hydrophobicity of petroleum hydrocarbon degrading Burkholderia strains and their interactions with NAPLs and surfaces Colloids and Surfaces B. Biointerfaces 78:101-108
7. Chao A (1984) Non-parametric estimation of the number of classes in a population. Scand J Stat 11:783-791

8. Chao AMC, Ma , Yang MCK (1993) Stopping rules and estimation for recapture debugging with unequal failure rates. Biometrika 80:193-201
9. Chien CC, Chen CC, Choi MH, Kung SS, Wei YH (2007) Production of poly-hydroxybutyrate (PHB) by *Vibrio* spp. isolated from marine environment. Journal of Biotechnology 132:259-263
10. Coleman NV, Bui NB, Holmes AJ (2006) Soluble di-iron monooxygenase gene diversity in soils, sediments and ethane enrichments. Environmental Microbiology 8:1228-1239 |
11. Cooley RB, Dubbels BL, Sayavedra-Soto LA, Bottomley PJ, Arp DJ (2009) Kinetic characterization of the soluble butane monooxygenase from *Thauera butanivorans*, formerly *'Pseudomonas butanovora'*. Microbiology 155:2086-2096
12. Davis JB, Updegraff DM Microbiology in the Petroleum Industry. Magnolia Petroleum Company. 18: Field Research Laboratories, Dallas, Texas, pp 215-238
13. Dimitriu PA, Pinkart HC, Peyton BM, Mormile MR (2008) Microbial diversity of a meromictic soda lake in Washington, USA: spatial and temporal. Applied and Environmental Microbiology 74:4877-4888
14. Ewing B, Hillier LD, Wendl MC, Green P (1998) Basecalling of automated sequencer traces using phred. I. Accuracy assessment. Genome Research 8:175-185
15. Fierer N, Bradford MA, Jackson RB (2007) Toward an Ecological Classification of Soil Bacteria. Ecology 88:1354-136
16. Galand PE, Casamayor EO, Kirchman DL, Lovejoy C (2009) Ecology of the rare microbial biosphere of the Arctic Ocean. Proc Natl Acad Sci USA 106:22427-22432
17. Ginkel CG, Welten HGJ, De Bont JAM (1987) Oxidation of Gaseous and Volatile Hydrocarbons by Selected Alkene-Utilizing Bacteria. Applied and Environmental Microbiology 53:2903-2907
18. Gomez-Leon J, Villamil L, Lemos ML, Novoa B, Figueras A (2005) Isolation of *Vibrio alginolyticus*and *Vibrio splendidus* from Aquacultured Carpet Shell Clam (*Ruditapes decussatus*) Larvae Associated with Mass Mortalities. Applied and Environmental Microbiology 71:98-104

19. Guan TW, Liu Y, Zhao K, Xia ZF, Zhang XP, Zhang LL (2010) *Actinopolyspora xinjiangensis* sp. nov., a novel extremely halophilic actinomycete isolated from a salt lake in Xinjiang, China. Antonie van Leeuwenhoek 98:447-453
20. Hartman M, Lee S, Hallam SJ, Mohn WW (2009) Bacterial, archaeal and eukaryal community structures throughout soil horizons of harvested and naturally disturbed forest stands. Environmental Microbiology 11:3045-3062
21. Hatzenpichler R, Lebedeva EV, Spieck E, Stoecker K, Richter A, Daims H, Wagner M (2007) A moderately thermophilic ammonia-oxidizing crenarchaeote from a hot spring. PNAS 105:2134-2139
22. Hennessee CT, Seo JS, Alvarez AM, Li QX (2009) Polycyclic aromatic hydrocarbon-degrading species isolated from Hawaiian soils: *Mycobacterium crocinum* sp. nov., *Mycobacterium pallens* sp. nov., *Mycobacterium rutilum* sp. nov., *Mycobacterium rufum* sp. nov. and *Mycobacterium aromaticivorans* sp. Nov Int J Syst Evol Microbiol 59:378-387
23. Hitzman DC, Schumacher D, Clavareau L (2009) Strategies for surface geochemical surveys in southeast Asia: Best practice designs and recent case studies. Proceedings, Indonesian Petroleum Association, Thirty-Third Annual Convention & Exhibition.
24. Holmes AJ, Coleman NV (2008) Evolutionary ecology and multidisciplinary approaches to prospecting for monooxygenases as biocatalysts. Antonie van Leeuwenhoek 94:75-84
25. Horvitz L (1985) Geochemical Exploration for Petroleum. Science 229:821-827
26. Janssen PH (2006) Identifying the Dominant Soil Bacterial taxa in Libraries. of 16S rRNA and 16S rRNA Genes. Applied and Environmental Microbiology 72:1719-1728
27. Kao CM, Chen CS, Tsa FY, Yang KH, Chien CC, Liang SH, Yang C, Chen SC (2010) Application of real-time PCR, DGGE fingerprinting, and culture-based method to evaluate the effectiveness of intrinsic bioremediation on the control of petroleum-hydrocarbon plume. Journal of Hazardous Materials 178:409-416

28. Kimura M (1980) A simple method for estimating evolutionary rates of base substitutions through comparative studies of nucleotide sequences. Journal of Molecular Evolution 16:111-20
29. Kniemeyer O, Musat F, Sievert SM, Knittel K, Wilkes H, Blumenberg M, Michaelis W, Classen A, Bolm C, Joye SB, Widdel F (2007) Anaerobic oxidation of short-chain hydrocarbons by marine sulphate-reducing bacteria. Nature 449:898-901
30. Kotani T, Kawashima Y, Yurimoto H, Kato N, Sakai Y (2006) Gene structure and regulation of alkane monooxygenases in propane-utilizing *Mycobacterium* sp. TY-6 and *Pseudonocardia* sp. TY-7. Journal of Bioscience and Bioengineering 102:184-192
31. Krause L, Diaz NN, Edwards RA, Gartemann KH, Krömeke H, Neuweger H, Pühler A, Runte KJ, Schlüter A, Stoye J, Szczepanowski R, Tauch A, Goesmann A (2008) Taxonomic composition and gene content of a methane-producing microbial community isolated from a biogas. Journal of Biotechnology 136:91-101
32. Krober M, Bekel T, Diaz NN, Goesmann A, Jaenicke S, Krause L, Millera D, Runte KJ, Viehöverc P, Pühlera A, Schlütera A (2009) Phylogenetic characterization of a biogas plant microbial community integrating clone library 16S-rDNA sequences and metagenome sequence data obtained by 454-pyrosequencing. Journal of Biotechnology 142:38-49
33. Kudo Y, Nakajima T, Miyaki T, Oyazu H (1997) Methanogen Flora of Paddy Soils in Japan. FEMS Microbiology Ecology 22:39-48
34. Kumar S, Tamura K, Jakobsen IB, Nei M (2001) MEGA2: Molecular Evolutionary Genetics Analysis software. Bioinformatics 17:1244-1245
35. Lacoste A, Jalabert F, Malham SK, Cueff A, Poulet SA (2001) Stress and Stress-Induced Neuroendocrine Changes Increase the Susceptibility of Juvenile Oysters (*Crassostrea gigas*) to*Vibrio splendidus*. Applied and Environmental Microbiology. 2304-2309
36. Lane DJ (1991) 16S/23S rRNA sequencing. In: Stackebrandt E, Goodfellow M (eds) Nucleic Acid Techniques in Bacterial Systematics, John Willey & Sons, Chinchester. pp 115-175

37. Leahy JG, Batchelor PJ, Morcomb SM (2003) Evolution of the soluble diiron monooxygenases. FEMS Microbiology Reviews 27:449-479

38. Li H, Yang SZ, Mu BZ, Rong ZF, Zhang J (2007) Molecular phylogenetic diversity of the microbial community associated with a high-temperature petroleum reservoir at an offshore oilfield. FEMS Microbiology Ecology 60:74-84

39. Lovatini A, Myers K, Watterson P, Campbell T (2010) An integrated approach to exploration data in the *Potiguar* basin, offshore Brazil. First Break 28:55-59

40. Magurran AE (2004) Measuring Biological Diversity. Blackwell Publishing, Oxford.

41. Mendez MO, Neilson JW, Maier RM (2008) Characterization of a Bacterial Community in an Abandoned Semiarid Lead-Zinc Mine Tailing Site. Applied and Environmental Microbiology 74:3899-3907

42. Pimentel-Elardo SM, Tiro LP, Grozdanov L, Hentschel U (2008) *Saccharopolyspora cebuensis* sp. nov., a novel actinomycete isolated from a Philippine sponge (Porifera). International Journal of Systematic and Evolutionary Microbiology 58:628-632

43. Pukall R, Gehrich-Schröter G, Lapidus A, Nolan M, Del Rio TG, Lucas S, Chen F, Tice H, Pitluck S, Cheng JF, Copeland A, Saunders E, Brettin T, Detter JC, Bruce D, Goodwin L, Pati A, Ivanova N, Mavromatis K, Ovchinnikova G, Chen A, Palaniappan K, Land M, Hauser L, Chang YJ, Jeffries CD, Chain P, Göker M, Bristow J, Eisen JA, Markowitz V, Hugenholtz P, Kyrpides NC, Klenk HP, Han C (2009) Complete genome sequence of *Jonesia denitrificans* type strain (Prevot 55134^{T}). Standards in Genomic Sciences 1:3

44. Qin S, Chen HH, Klenk HP, Kim CJ, Xu LH, Li WJ (2010) *Saccharopolyspora gloriosae* sp. nov., an endophytic actinomycete isolated from the stem of *Gloriosa superba* L. Int J Syst Evol Microbiol 60:1147-1151

45. Quaiser A, Ochsenreiter T, Lanz C, Schuster SC, Treusch AH, Eck J, Schleper C (2003) Acidobacteria form a coherent but highly diverse group within the bacterial domain: evidence from environmental genomics. Molecular Microbiology 50:563-575

46. Rasheed MA, Prasanna MV, Kumar TS, Patil DJ, Dayal AM (2008) Geo-microbial prospecting method for hydrocarbon exploration in Vengannapalli Village, Cuddapah Basin, India. Current Science 95:361-366

47. Schloss PD, Handelsman J (2005) Introducing DOTUR, a Computer Program for Defining Operational Taxonomic Units and Estimating Species Richness. Applied and Environmental Microbiology 71:1501-1506

48. Schumacher D (2000) Surface geochemical exploration for oil and gas: New life for an old technology. In: Geo-Microbial Technologies. The Leading Edge. pp 258-261

49. Shennan JL (2006) Utilisation of C2-C4 gaseous hydrocarbons and isoprene by microorganisms. Journal of Chemical Technology and Biotechnology 81:237-256

50. Silva CC, Ederson CJ, Torres APR, Sousa MP, Santiago VMJ, Oliveira VM (2010) Investigation of Bacterial Diversity in Membrane Bioreactor and Conventional Activated Sludge Processes from Petroleum Refineries Using Phylogenetic and Statistical Approaches. J Microbiol Biotechnol 20:447-459

51. Sogin ML, Morrison HG, Huber JA, Welch DM, Huse SM, Neal PR, Arrieta JM, Herndl GJ (2006) Microbial diversity in the deep sea and the underexplored "rare biosphere". Proc Natl Acad Sci USA 103:12115-12120

52. Spang A, Hatzenpichler R, Brochier-Armanet C, Rattei T, Tischler P, Spieck E, Streit W, Stahl DA, Wagner M, Schleper C (2010) Distinct gene set in two different lineages of ammonia-oxidizing archaea supports the phylum Thaumarchaeota. Trends in Microbiology 18:331-340

53. Suthindhiran K, Kannabiran K (2009) Cytotoxic and Antimicrobial Potential of Actinomycete Species*Saccharopolyspora salina* VITSDK4 isolated from the Bay of Bengal Coast of India. American Journal of Infectious Diseases 5:90-98

54. Tang SK, Wang Y, Cai M, Zhi XY, Lou K, Xu LH, Jiang CL, Li WJ (2009) *Saccharopolyspora halophila*sp. nov., a novel halophilic actinomycete isolated from a saline lake in China. Int J Syst Evol Microbiol 59:555-558

55. Thomas SH, Wagner RD, Arakaki AK, Skolnick J, Kirby JR, Shimkets LJ, Sanford RA, Löffler FE (2008) The mosaic genome of *Anaeromyxobacter dehalogenans* strain 2CP-C suggests an aerobic common ancestor to the delta-proteobacteria. PLoS One 3:e2103
56. Thompson JD, Gibson TJ, Plewniak F, Jeanmougin F, Higgins DG (1997) The ClustalX windows interface: flexible strategies for multiple sequence alignment aided by quality analysis tools. Nucleic Acids Research 24:4876-4882
57. Tucker J, Hitzman D (1996) Long-term and seasonal trends in the response of hydrocarbon-utilizing microbes to light hydrocarbon gases in shallow soils. In: Schumacher D, Abrams MA (eds) Hydrocarbon migration and its near-surface expression: AAPG Memoir, vol 66.353-357
58. Venier P, Varotto L, Rosani U, Millino C, Celegato C, Bernante F, Lanfranchi G, Novoa B, Roch P, Figueras A, Pallavicini A (2011) Insights into the innate immunity of the Mediterranean mussel*Mytilus galloprovincialis*. BMC Genomics 12:69
59. Vomberg A, Klinner U (2000) Distribution of alkB genes within n-alkane-degrading bacteria. Journal of Applied Microbiology 89:339-348
60. Wagner (2002) Case Histories of Microbial Prospection for Oil and Gas, Onshore and offshore in Northwest Europe, in Surface exploration case histories: Applications of geochemistry, magnetics, and remote sensing. SEG Geophysical References Series 11:453-479
61. Yassin AF (2009) *Saccharopolyspora rosea* sp. nov., isolated from a patient with bronchial carcinoma. Int J Syst Evol Microbiol 59:1148-1152
62. Yi H, Chun J (2004) *Nocardioides aestuarii* sp. nov., isolated from tidal flat sediment. International Journal of Systematic and Evolutionary Microbiology 54:2151-2154
63. Youssef N, Elshahed MS (2009) Diversity rankings among bacterial lineages in soil. The ISME Journal 3:305-313
64. Yuan LJ, Zhang YQ, Guan Y, Wei YZ, Li QP, Yu LY, Li WJ, Zhang YQ (2008) *Saccharopolyspora antimicrobica* sp. nov., an actinomycete from soil. Int J Syst Evol Microbiol 58:1180-1185

65. Zhang F, She Y, Zheng Y, Zhou Z, Kong S, Hou D (2010) Molecular biologic techniques applied to the microbial prospecting of oil and gas in the Ban 876 gas and oil field in China. Applied Microbiology and Biotechnology 86:1183-1194

Chapter 7

Encapsulation of Docetaxel in Oily Core Polyester Nanocapsules Intended for Breast Cancer Therapy

Ibrahima Youm[1], Xiaoyan Yang[1], James B Murowchick[2], and Bi-Botti C Youan[1]

[1]Laboratory of Future Nanomedicines and Theoretical Chronopharmaceutics, Division of Pharmaceutical Sciences, University of Missouri-Kansas City, 2464 Charlotte Street, Kansas City, USA

[2]Department of Geosciences, University of Missouri-Kansas City, 420 Flarsheim Hall, 5110 Rockhill Rd., Kansas City, USA

ABSTRACT

This study is designed to test the hypothesis that docetaxel [Doc] containing oily core nanocapsules [NCs] could be successfully prepared with a high percentage encapsulation efficiency [EE%] and high drug loading. The oily core NCs were generated according to the emulsion solvent diffusion method using neutral Labrafac CC and poly (*d*, *l*-lactide) [PLA] as oily core and shell, respectively. The engineered NCs were characterized for particle mean diameter, zeta potential, EE%, drug release kinetics, morphology, crystallinity, and cytotoxicity on the SUM 225 breast cancer cell line by dynamic light scattering, high performance liquid chromatography, electron microscopies, powder X-ray diffraction, and lactate dehydrogenase bioassay. Typically, the formation of Doc-loaded, oily core, polyester-based NCs was evidenced by spherical nanometric particles (115 to 582 nm) with a low polydispersity index (< 0.05), high EE% (65% to 93%), high drug loading (up to 68.3%), and a smooth surface. Powder X-ray diffraction analysis revealed that Doc was not present in a crystalline state because it was dissolved within the NCs' oily core and the PLA shell. The drug/polymer interaction has been indeed thermodynamically explained using the Flory-Huggins interaction parameters. Doc release kinetic data over 144 h fitted very well with the Higuchi model ($R^2 > 0.93$), indicating that drug release occurred mainly by controlled diffusion. At the highest drug concentration (5 μM), the Doc-loaded oily core NCs (as a reservoir nanosystem) enhanced the native drug cytotoxicity. These data suggest that the oily core NCs are promising templates for controlled delivery of poorly water soluble chemotherapeutic agents, such as Doc.

BACKGROUND

In cancer therapy, most of the proposed formulations present certain drawbacks related to the formulation properties including low drug loading, toxicity, and unsuitable release pattern. An ideal formulation should provide biocompatible nanosized particles and high drug loading with sustained-release characteristics. This allows releasing the drug in the target site in its therapeutic concentration and preventing drug inefficiency and side effects. The current research was aimed to prepare highly loaded docetaxel [Doc] oily core nanocapsules [NCs].

Doc, a semisynthetic analog of paclitaxel, is an extract from the needles of the European yew tree *Taxus baccata* [1]. It is prepared by chemical modification of 10-deacetylbacattin III, an inactive precursor compound, and then isolated [2]. Doc is a highly potent, cytotoxic, and antimitotic agent used in the treatment of various types of cancers, including metastatic breast, ovarian, prostate, advanced non-small-cell lung, head/neck, and advanced gastric cancers by inhibiting the microtubule depolymerization of free tubulin [3, 4]. Due to its poor water solubility (10 to 20 μg/l), polysorbate 80 has been markedly used to improve the aqueous solubility of Doc [5-7]. This currently available, marketed formulation has been associated with the absence of selectivity for target tissues, serious dose limiting toxicities, and hypersensitivity reactions, as well as sensory and motor neuropathies that are sometimes severe and irreversible. Previously, various alternative formulations, including NCs, pegylated liposomes, targeted immunoliposomes, Doc-fibrinogen-coated olive oil droplets, and cyclodextrins [8-15] have been intensively developed for the delivery of Doc. However, the nanosized polymeric nanoparticles represent promising drug-delivery systems which have some advantages such as biodegradability, good biocompatibility, non-toxicity, higher stability, and controlled drug delivery. Polymeric nanoparticles (nanospheres and NCs) not only maintain a prolonged circulation time in the body (especially when pegylated) by avoiding the reticuloendothelial system, but also can extravagate and accumulate into the tumor tissue. This is likely due to the reliance of these nanoparticles on passive accumulation through enhanced permeability and retention, which is highly dependent on adequate blood flow to the tumor [16].

According to the literature, the NCs correspond to nanostructures with polymeric wall enveloping an oily core, whereas the nanospheres consist of a polymeric matrix [17]. There is increasing scientific evidence supporting the notion that certain lipids are able to inhibit both presystemic drug metabolism and P-glycoprotein-mediated drug efflux [18]. Several polymers have been proposed as nanocarriers for drug delivery systems. For example, poly (*d, l*-lactic acid) and poly (ε-caprolactone) [PCL] have been extensively used as nanocarriers because of their excellent biocompatibility and biodegradability. These polyesters have been approved by the US Food and Drug Administration and are the most widely used commercial polymers for drug delivery [19, 20]. Presently, the only available Doc formulations for clinical use consist of intravenous [IV] solutions containing Tween 80® (Sanofi-aventis, Bridgewater, NJ, USA). These solutions namely Taxotere® and Docetaxel®, 10 to 20 mg/ml, are administered IV at a dose ranging from 60 to 100 mg/m^2 over 1 h every 3 weeks [21]). However, such high doses and a long term medication schedule may produce more severe side effects [22]. One of the reasons could be ascribed to the composition of the formulation and the poor control of the drug release rate. Therefore, the development of compatible polymer-based nanocarriers with high drug loading might be a helpful subject in cancer research.

The current research was aimed to prepare highly loaded Doc oily core NCs. However, the relationship between drug and formulation excipients has also been investigated to better control the encapsulation process.

MATERIALS AND METHODS

Materials

Poly(*d, l*-lactide) [PLA], Resomer® R206, Mw 125 kDa, with an inherent viscosity of 1.0 dl/g; PLA, Resomer® R207, Mw 209 kDa, with an inherent viscosity of 1.5 dl/g; and PLA, Resomer® R208, Mw 250 kDa, with an inherent viscosity of 1.8 dl/g were purchased from Boehringer Inc. (Ridgefield, CT, USA). PCL (Mw = 72 kDa) was kindly provided by Union Carbide (Danbury, CT, USA). Polyvinyl alcohol [PVA] (9 to 10 kDa and 30 to 70 kDa) and ethyl acetate were obtained from Sigma-Aldrich (St. Louis, MO, USA). Docetaxel or Doc was purchased from LC Laboratories (Woburn, MA, USA). Labrafac CC (caprylic/capric triglyceride, $d = 0.945$ g/cm^3) was kindly supplied by Gattefosse Corporation (St-Priest, France) as a gift. For the high-performance liquid chromatography [HPLC] analysis, acetonitrile and methanol were supplied from Fisher (Thermo Fisher Scientific, Fair Lawn, NJ, USA). All the other reagents were of analytical grade and used without further purification.

Preparation of the Nanocapsules

Experimental Design and General Procedure for Doc-loaded Nanocapsules

Fifteen formulations of Doc-loaded NCs were prepared by emulsion-diffusion method as previously described [19]. Briefly, the polyester (40 to 360 mg) was solubilized in water-saturated ethyl acetate (10 ml); then, the neutral oil (0.1 to 0.9 ml) containing Doc (2 to 18 mg) was further added to the organic mixture. The resulting solution was emulsified in 40 ml of PVA 2.5% to 5% (*w/v*) aqueous phase solution by homogenization (homogenizer, IKA ULTRA-TURRAX T-25, IKA Labortechnik, Staufen, Germany) at 8,000 rpm for 10 min. A large volume (200 ml) of deionized water was added dropwise into the previous solution to promote the diffusion of ethyl acetate into the aqueous phase. To remove the organic solvent, the nanosuspension was stirred under vacuum at 40°C Rotavapor®RII (BUCHI Labortechnik AG, Flawil, Switzerland) for 30 min. NCs were recovered by ultracentrifugation at 12,000 rpm at 5°C for 30 min, washed twice with deionized water to remove the excess of PVA, and freeze-dried for 12 h (Labconco Corp. Kansas City, MO, USA). Blank NCs were prepared without Doc using the above method. The composition of the 15 formulations is listed in the first four columns of Table S1 in.

Screening for Polyester Selection Based on the Blank Particle Mean Diameter

For the first screening test, a set of four biodegradable polymers, including PLA R206, PLA R207, PLA R208, and PCL, was used to obtain a suitable formulation based on the mean diameter of the obtained blank NCs. An amount of 200 mg of polyesters was solubilized in an organic phase containing 10 ml of water-saturated ethyl acetate and 0.5 ml of neutral oil. Each experiment was performed in triplicate.

Screening for Stabilizer Selection Based on the Blank Particle Mean Diameter

Since PVA could be used as an emulsion stabilizer in the NCs' formulation process, it was important to first investigate the effect of PVA molecular weight (9 to 10 kDa and 30 to 70 kDa) on the blank NCs' mean diameter. Secondly, the influence of the PVA concentration (2.5% to 5%) on the particle mean diameter and polydispersity index [PDI] was studied.

Physicochemical Characterization of Docetaxel-loaded Nanocapsules

Particles' Mean Diameter and Zeta Potential Analysis

The particle mean diameter and PDI of the Doc-loaded NCs were measured by dynamic light scattering [DLS] (Zetasizer Nano ZS series from Malvern Instruments Ltd., Worcestershire, UK) as recently reported [23]. The sample to be measured was appropriately diluted with water and briefly sonicated for 2 min. The measurement of each sample was completed at a scattering angle of 175°. Each measurement was done in triplicate, and the average effective diameter and polydispersity were recorded.

Drug Loading and Percent Encapsulation Efficiency of Docetaxel-loaded Nanocapsules

The freeze-dried NCs (2 mg) were dissolved in 0.1 ml of dichloromethane and diluted in methanol at the ratio of 1:14 *v/v*. After suitable dilutions, the amount of the encapsulated Doc was determined by HPLC (Waters Corporation, Milford, MA, USA) equipped with an UV detector at 230 nm. Isocratic flow of the mobile phase, composed of methanol/water/acetonitrile (30:30:40 *v/v/v*), was employed at a flow rate of 1.0 ml·min^{-1} with a 10-µl injection volume. Doc separation was completed using an XB ridge™ column C-18, at 4.6 × 150 mm and 3.5 µm (Waters Corporation, Milford, Massachusetts). The experimental Doc loading was quantified using the peak area of each NC formulation. Drug loading [DL] and percent encapsulation efficiency [EE%] were calculated according to Equations 1 and 2, respectively:

$$DL = \frac{\text{Drug amount}}{\text{Polymer amount} + \text{drug amount}} \times 100\% \tag{1}$$

$$EE = \frac{\text{Experimental drug loading}}{\text{Theoretical drug loading}} \times 100\% \tag{2}$$

Scanning Electron Microscopy Analysis

Scanning electron microscopy [SEM] was used to assess the NCs' morphology. A droplet of NCs' suspension was put into a grid. The excess of the fluid was removed by wicking it off with an adsorbent paper, and then, it was visualized under a Hitachi S4700 cold-cathode field emission SEM [FESEM] (Hitachi High-Technologies Corporation, Minato-ku, Tokyo, Japan). The particles were sprinkled onto a stub covered with an adhesive conductive carbon tab, then sputter-coated with a fine layer of platinum metal. Then, the particles were imaged in the FESEM at 2 to 5 kV.

Transmission Electron Microscopy Analysis

The selected samples were examined with a JEOL 1400 transmission electron microscope [TEM] (JEOL Ltd., Tokyo, Japan) and photographed digitally on a Gatan axis-mount 2 k × 2 k digital camera (Gatan, Inc., Pleasanton, CA, USA). The freeze-dried samples were put into a small mold, referred to as a BEEM capsule, and then imbedded in liquid epoxy resin (Epon-Araldite, Sigma-Aldrich Corporation, St. Louis, MO, USA). The resin was polymerized at 60°C for 2 days, and then, ultrathin 80-nm sections were cut on a Leica UCT ultramicrotome (Leica

Microsystems Ltd., Milton Keynes, UK) with a Diatome diamond knife (Diatome, Hatfield, PA, USA). The sections were collected on 200-mesh copper grids and put into the TEM for imaging on a Gatan digital camera.

Powder X-ray Diffraction Pattern Analysis

Powder X-ray diffraction [PXRD] analysis of the freeze-dried NCs was performed using a MiniFlex automated X-ray diffractometer (Rigaku, The Woodlands, TX, and USA) at room temperature. Ni-filtered Cu Kα radiation was used at 30 kV and 15 mA. The diffraction angle covered from $2\theta = 5°$ to $2\theta = 60°$, with a step size of 0.05°/step and a count time of 3 s/step (effectively 1°/min). The diffraction patterns were processed using Jade 8+ software (Materials Data, Inc., Livermore, CA, USA).

Refractive Index Measurement

The oil, water-saturated ethyl acetate, and ethyl acetate-saturated water refractive index values were measured experimentally at 25°C (Auto Abbe 10500 Refractometer; Reichert Analytical Instruments, Depew, NY, USA) using Milli-Q water (Millipore Co., Billerica, MA, USA) as a reference. A droplet of liquid was deposited on the prism surface. The obtained values are the average of five measurements.

In Vitro Drug Release Kinetics

This experiment was performed using the equilibrium dialysis method for 144 h. Specifically, a known amount of Doc-loaded oily core NCs (1 mg) was suspended in a dialysis bag (Spectra/Float-A-Lyzer, MWCO 3.5-5 kDa, Spectrum Laboratories Inc. Rancho Dominguez, CA, USA) containing 5 ml of phosphate buffer saline [PBS] (Sigma-Aldrich, St. Louis, MO, USA). The bag containing the NCs' suspension was placed in 40 ml of PBS. The system was placed in a shaking water bath (BS-06, Lab. Companion, Des Plaines, IL, USA) at 37°C with an agitation speed of 50 rpm. At predetermined time intervals, 500 µl of PBS solution was withdrawn from the immersion medium and replaced by the same volume of fresh medium. The cumulative percentage of drug released for each time point was calculated as a percentage of the total drug loading of the NCs tested. The quantitative analysis of the data obtained from the study was confirmed using Higuchi›s kinetic model [24] to elucidate the mechanism of the drug release.

Solubility Parameter Determination

To predict the compatibility between the Doc (solubilizate) and polymer (solvent), the Flory-Huggins solubility parameter was evaluated [χ_{sp}]. Our goal was to determine the interaction parameter χ_{sp} [25] using the thermodynamic approach based on the extended Hildebrand solubility. According to Hildebrand, the solubility parameter [δ], defined as the square root of the cohesive energy density [CED], is equal to the energy of vaporization ΔE_v per unit of molar volume (Equation 3) [26]. The solubility parameter is used to calculate χ_{sp} using Equation 3:

$$\delta = (\mathrm{CED})^{1/2} = \left(\Delta E_{\mathrm{v}}/V_{\mathrm{m}}\right)^{1/2} \tag{3}$$

Hansen modified the Hildebrand approach and divided δ into three components that take into account the force of the dispersion [δ_d], the polarity [δ_p], and the hydrogen bonds [δ_h]. Therefore, δ is calculated using Equation 4:

$$\delta_t^2 = \delta_d^2 + \delta_p^2 + \delta_h^2 \tag{4}$$

Equation 5 below was used to estimate the total Hildebrand solubility parameter based on the refractive index [n_D], where δ_t was the Hildebrand solubility parameter in $(\mathrm{cal/cm^3})^{1/2}$, and 304.5 was an empirically determined constant [27]. The refractometer was calibrated using pure water according to the instrument manual. Then, the total solubility parameter was calculated by using Equation 5:

$$\delta_t = \sqrt{304.5 \left(\frac{n_D^2 - 1}{n_D^2 + 2} \right)}$$

(5)

An interaction parameter with a value smaller than 0.5 (i.e., $\chi_{sp} < 0.5$) indicates that the solvent polymers are compatible. The interaction parameter can be calculated from the solubility parameters [28]. Considering the corresponding δ_t for each component, the value of the interaction parameter (χ_{sp}) can be estimated from the Hildebrand solubility parameters δ_s and δ_p (for solubilisant and polymer, respectively). On the basis of regular solution theory, the relationship between the Flory-Huggins interaction parameter and the solubility parameter is defined by using Equation 6:

$$\chi_{sp} = (\delta_s - \delta_p)^2 . V_m / R . T$$

(6)

Where V_m is the molar volume of the drug, R is the ideal gas constant (8.314 $J \cdot K^{-1} \cdot mol^{-1}$), and T is the temperature in Kelvin (293.15 K) [29].

In Vitro Evaluation of Doc-loaded Nanocapsules' Cytotoxicity

The NCs' cytotoxicity was evaluated using the SUM 225 cell line (Asterand, Inc., Detroit, MI, USA). The cytotoxicity after treatment of these cells with native Doc and Doc-loaded NCs at different equivalent drug concentrations (1 nM, 2.5 μM, and 5 μM) was evaluated by the lactate dehydrogenase [LDH] assay. The cells were plated at a density of 5×10^3 cells per well for 24 h in 96-well plates in a standard growth medium prior to exposure to the above materials. Blank NCs, and medium containing 0.5% dimethyl sulfoxide [DMSO] are used as negative controls, while Triton-X 1% (Sigma-Aldrich Corporation, St.

Louis, MO) was used as a positive control and incubated for 24 h at 37°C in 5% CO_2. After the treatment, the LDH reagents were used according to the manufacturer's instruction (Promega Life Sciences, Madison, WI, USA). The experimental results were expressed as mean values of six measurements ($n = 6$), and the cytotoxicity was calculated by the following formula:

$$\text{Cytotoxicity (\%)} = \frac{\text{Experimental} - \text{Background}}{\text{Positive} - \text{Background}} \times 100 \tag{7}$$

Where, experimental, background, and positive represent the fluorescence intensity of NC-treated wells, background wells (wells without cells), and positive control wells (cells treated with 1% of Triton X-100), respectively. The fluorescence intensity was detected by using a microplate reader (DTX 800 multimode microplate reader, Beckman Coulter, Brea, CA, USA) at an excitation wavelength of 560 nm and emission wavelength of 590 nm.

RESULTS AND DISCUSSION

Polymer Selection Based on the Mean Diameter of Blank Nanocapsules

This experiment was performed to find out the suitable polymer using small-sized NCs. Four polymers were initially screened including PLA R206, PLA R207, PLA R208, and PCL. Figure 1 shows the effect of different biodegradable polymers on the average diameter of blank NCs. The NCs' mean diameters were 127.5 ± 19.2 nm for NC-PLA206, 123.8 ± 0.9 nm for NC-PLA207, 110.8 ± 8.6 nm for NC-PLA208, and 124.6 ± 3.1 nm for NC-PCL. These findings indicated a statistically significant decrease of PLA NCs› diameters with increasing polymer molecular weights ($P < 0.003$, T test). Based on their smaller mean diameter, the NCs prepared with PLA R208 were selected for the subsequent studies. The results did not show any difference regarding the zeta potential values ($\zeta = -36.5 \pm 9$ mV) among the batches of

NCs (data not shown). This high potential value also contributed to the stabilization of the nanosuspension.

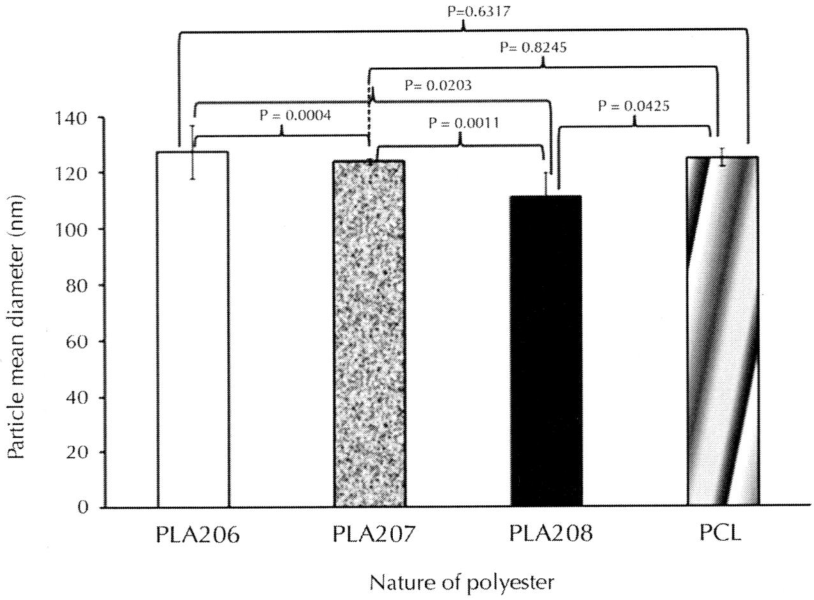

Figure 1: Influence of the nature of biodegradable polyester on blank oily core nanocapsules mean diameter. $n = 3$; S.D.: standard deviation between the three assays: PLA206 (105 kDa), PLA207 (209 kDa), PLA208 (250 kDa), and PCL (72 kDa).

Stabilizer Concentration and Molecular Weight Effects on the Blank Particle Mean Diameter

This preliminary experiment was performed to select the accurate PVA molecular weight and concentration with the goal of minimizing the particle size. Figure 2A, B, C, D shows that the NCs' mean diameter decreased with increasing both the PVA's molecular weight (9 to 10 kDa, to 30 to 70 kDa) and concentration (2.5% to 5%, *w/w*). Oppositely, previous studies have pointed out the increase of the PLGA NP mean diameter when the concentration of PVA was increased from 2% to 6% [30]. This effect was attributed to the increase of the external phase viscosity, which decreases the molecular diffusion rate and Ostwald ripening phenomenon. This revealed that the PVA concentration is not the only parameter governing the particle mean diameter. According to the overall results, PLA 208 and PVA (30 to 70 kDa, 5%) were finally selected for the following NC preparation.

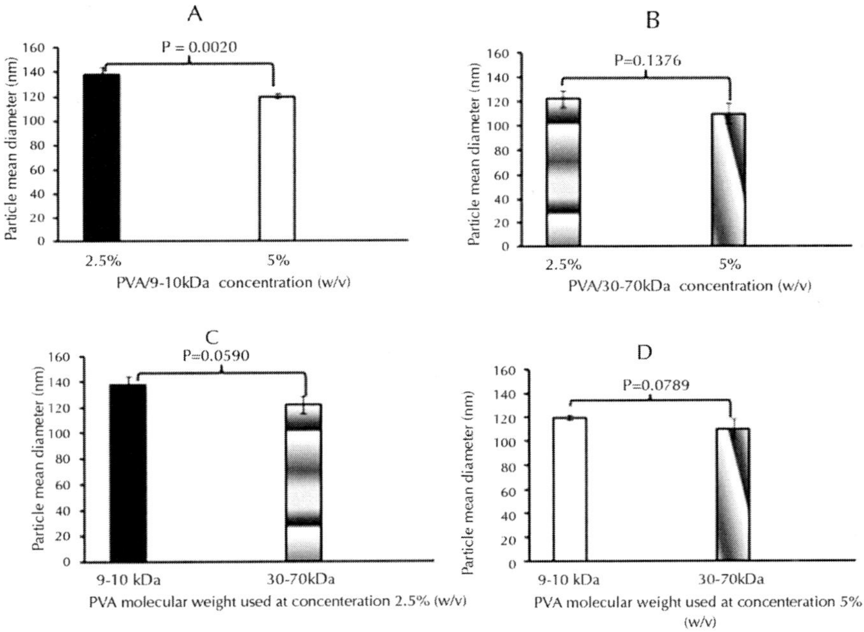

Figure 2: Influence of PVA concentrations (A, B) and molecular weights (C, D) on the blank NC mean diameter. This is detected by DLS ($n = 3$; S.D.: standard deviation between the three assays).

Preparation and Characterization of Docetaxel-loaded Nanocapsules

For optimization purposes, 15 batches of PLA R208 Doc-loaded oily core NCs were prepared using the above method. As shown in Table S1 in, the lowest value of the NCs' diameter was obtained with F_6 (115.6 nm), while the largest particle diameter was obtained with F_9 (582.8 nm). The PDI ranged from 0.004 to 0.318 (F_{14} and F_9, respectively).

It was found that the particle mean diameter is strongly dependent to the polymer amount. Indeed, at a low PLA amount (40 mg), the particle mean diameter increases with increasing drug amount (see F_9 versus F_{10}, $P < 0.0001$). This might be explained by the fact that the lipophilic feature tends to decrease the leakage of the drug into the external aqueous medium, leading to improved drug content in the nanoparticles (which is consistent with the previous report) [19]. However, at a high PLA amount (360 mg), the particle mean diameter decreases with increasing drug amount (see F_4 versus F_5, $P < 0.0001$). The latter was not consistent with the commonly published report and might be a result of drug solubilization in the polymer matrix, leading to decrease the particle mean diameter. Thus, the solubilization capacity of PLA has a great importance in the preparation of the Doc-loaded NCs.

From Table S1 in 1, the data suggest that the NCs' mean diameter decreases with increasing oil content from 189 nm to 133 nm (see F_3 versus F_{13}, $P < 0.0001$). The results also show that at low oil content, the drug level did not have any effect on the NCs' mean diameter, which is consistent with the previous study [31].

From Table S1 in, it appears that for most of the formulations, the PDI value was less than 0.05, indicating a monodispersity according to the National Institute of Standard [32]. However, the polydispersity seems to be increased with decreasing PLA amount (see F_{13} versus F_1, from 0.005 to 0.296, $P = 0.0078$). This suggests that the PLA amount may contribute to ensure the NCs' monodispersity.

Table S1 in list also the EE% of the Doc-loaded NCs. Interestingly enough, the results showed that the EE% was mainly governed by the PLA content. A high PLA content led to a high EE% (see pairwise comparison: F_1 to F_{13}, F_3 to F_9, F_4 to F_{11}, F_5 to F_{10}, respectively). At a low PLA content, the EE% was decreasing regardless of the oil content (comparing F_9 to F_{10}, $P = 0.0006$). At a high oil content, a medium content of PLA is at least required to obtain a high EE% (see F_7 and F_{13}, $P = 0.9038$). The lowest EE% (F_9, 65.3%) was obtained when the PLA and oil contents were both at the lowest level (40 mg and 0.1 ml, respectively). This is because the encapsulation process of hydrophobic drugs into these particles results from the interaction between the drug, polymers, and oil. Thus, the drug loading and EE% were found to depend on its solubility in the polymeric material, which is strongly related to the polymer composition, its molecular weight, the drug and polymer interaction, and the presence of end-functional ester or carboxyl groups [31]. These findings were consistent with a previous report [19]. Once the NCs' physicochemical properties have been analyzed, their size and morphology can be most directly monitored by various forms of electron microscopy.

Physicochemical Characterization of Docetaxel-loaded Nanocapsules

Morphological Analysis

The particle mean diameter and morphology were analyzed by SEM and TEM. Figure 3 shows a typical SEM picture of spherical NCs with smooth surfaces and undetectable free drug crystals. The NCs' size as estimated by SEM correlated well with the size measured by the DLS showing particles in a nanometric size range. The TEM analysis shows clearly a white and shiny oily core where Doc was well dissolved [33] (Figure 4). Therefore, it is necessary to investigate the drug structure inside the NCs.

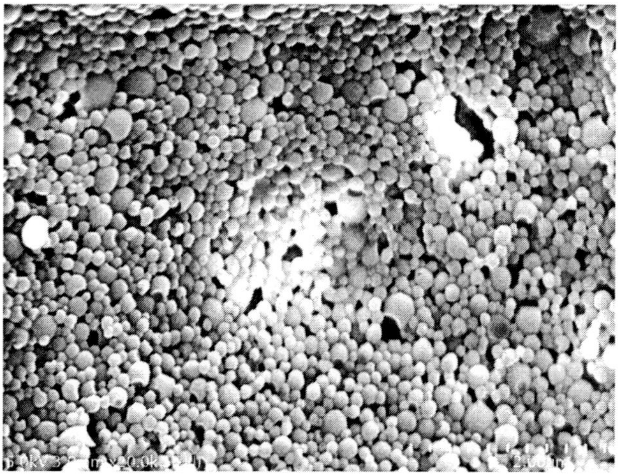

Figure 3: Scanning electron micrographs of Doc-loaded oily core nanocapsules after freeze drying (F6). Scale bar represents 0.2 μm.

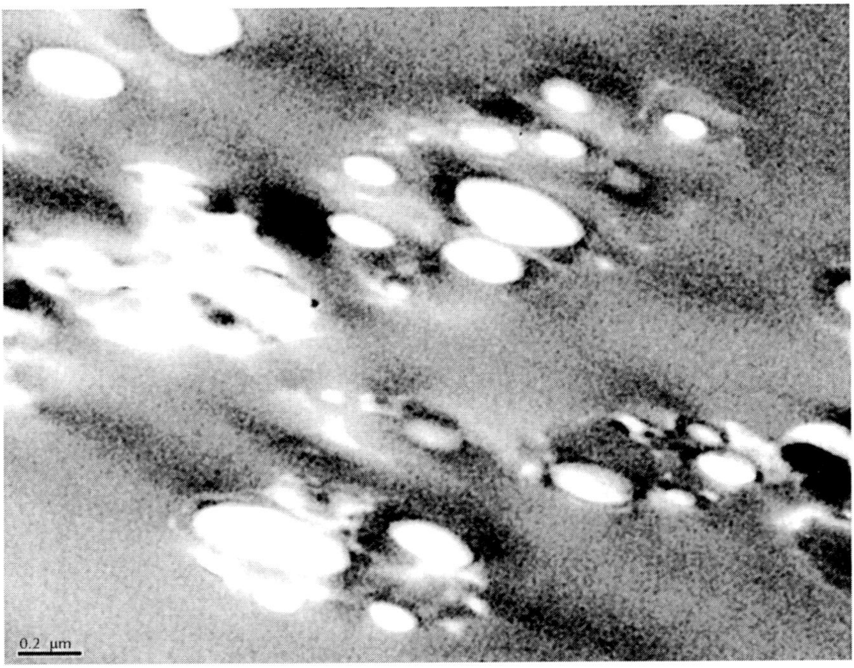

Figure 4: Transmission electron micrographs of PLA 208 oily core nanocapsules after freeze drying (F6). Scale bar represents 0.2 µm.

Powder X-ray Diffraction Analysis

This experiment aimed to characterize the crystallinity of Doc in the formulated NCs. Figure 5 shows the PXRD results of the formulations containing different amounts of Doc (F_6, F_{12}, and F_{13}) and the individual native chemical compounds used in the tested formulations. The native Doc exhibited sharp and characteristic diffraction peaks at 8.72°, 10.28°, and 11.7°, which is consistent with a previous report [34]. The Doc crystals are orthorhombic. The unit cell parameters (at room temperature) are a = 39.9345 Å, b = 12.7749 Å, c = 8.6644 Å, with V = 4420.2 Å. The number of motifs (Z) per cell is 4, and the density of the crystal is 1.295 [35]. The native PLA is characterized by a broad diffraction peak, which was centered at 25°. The PXRD pattern of the native PVA shows a broad lump between 17° and 20° both in the native component and in the tested formulations, which suggests that the PVA molecule is present in the NCs. The previous study indicated that the OH groups of PVA was adsorbed on the surface of the NCs and can keep the NCs in a pseudo-hydrated state [36]. The diffraction patterns of the native PLA exhibit a broad diffraction peak from $2\theta =$ 16.7° to $2\theta = 35°$. These results indicate that Doc is not present as a crystalline state, but is probably dissolved within the NCs' core and shell while some residual PVA might be present on the surface of the NCs.

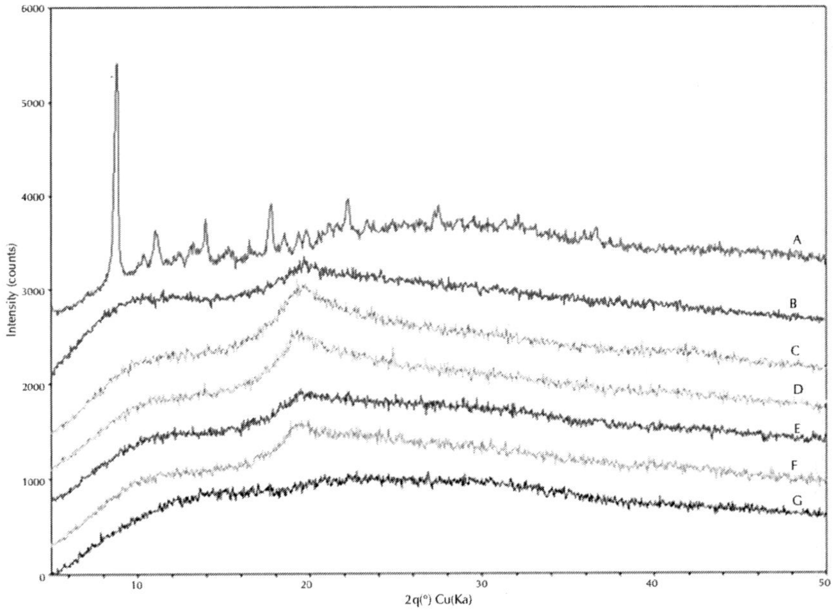

Figure 5: PXRD pattern. A, native Doc; B, F_{13} (PLA, 360 mg; Doc, 10 mg; oil, 0.9 ml); C, F_{12} (200 mg, 10 mg, 0.5 ml); D, F_6 (200 mg, 18 mg, 0.9 ml); E, blank nanocapsules; F,native PVA; and G, native PLA 208.

In Vitro Drug Release Study

The purpose of this study was to investigate the drug release mechanisms from Doc-loaded NCs. The cumulative percentages of Doc released from NCs as a function of time are reported in Figure 6. The results indicate that the drug release rate depends on the NCs' properties. The Doc oily core NCs was characterized by a sustained release profile (less than 40% of Doc is released within 60 h). This might be attributed to the hydrophobic interaction between the hydrophobic moiety of PLA, oil, and Doc. Compared to a previous result using nanosphere-based PLA [37], one can assume that the drug release is mainly due to the oil contained inside the NCs' core. The release data of Doc from the NCs (Figure 6) were fitted to the Higuchi model [38] to determine the drug release mechanism. The drug release constant (*k*) and regression coefficient (R^2) of the Higuchi model are shown in Table S2 in. Accordingly, the Doc release was best supported by Higuchi's model, i.e., based on Fickian diffusion, as it presented the highest values of linearity ($R^2 > 0.93$) for all formulations. From F_6, the release rate of Doc in 24 h corresponds to 0.9 nmol. This concentration is consistent with a previous report where the EC_{50} of Doc was ranged from 1 to 6.2 nmol. [39]. In these conditions, the low drug level in simulated plasma pH suggests that a triggering mechanism would be required to enhance drug release *in situ* within the targeted cancer cells in order to spare vital organs and significantly reduce the unwanted systemic side effect. To better understand the compatibility between Doc and PLA, as well as the neutral oil, the Flory-Huggins interaction parameter calculations were carried out as shown below.

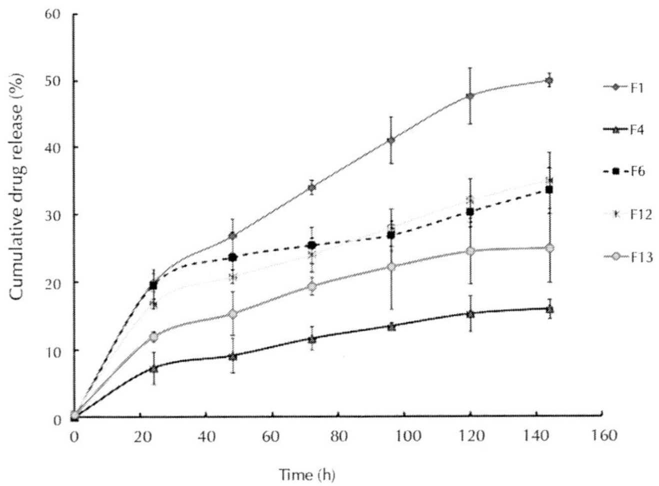

Figure 6: Cumulative drug release from Doc-loaded nanocapsules in PBS (pH 7.4), at 37 ± 0.5°C. Each point represents the mean value of three different experiments ± standard deviation.

Solubility Parameter Determination

To prove the suitability of PLA and oil for the optimal solubilization of Doc in the nanostructure, the Flory-Huggins interaction parameters were computed. The solubility parameters of Doc (δ_s) and PLA (δ_p) were calculated and listed in Table S3 in. According to the Flory-Huggins theory, the critical χ_{sp} value above which a polymer and a low molecular weight compound (i.e., drug) become miscible is < 0.5. Therefore, a lower value of χ_{sp} should result in a higher solubility. The obtained interaction parameters (χ_{sp}) between Doc and PLA and between Doc and oil were calculated. Table S3 in shows a low value of the interaction parameter (χ_{sp}) between Doc and PLA 0.65 (cal/cm^3)$^{1/2}$. In contrast, the interaction solubility value between Doc/oil was relatively high 1.64 (cal/cm^3)$^{1/2}$, indicating that Doc is more compatible with PLA than with the oil. This finding seems to show that a high PLA content may contribute to high drug EE% (up to 93%). To confirm the suitability of the oil/ethyl acetate mixture (phase A) with water (phase B) for the NCs' formation, the interfacial tension (γ_{AB}) between these two immiscible phases (A and B) was predicted by Kim and Burgess [40] using Equation 7:

$$\gamma_{AB} = \gamma_A - \gamma_B \ \exp\left(-\alpha V^{0.7}\right) + \gamma_B \tag{7.1}$$

Where γ_A and γ_B (72 mN/m at 25°C) are the surface tension of each phase (A or B), α is an exponential coefficient, and V is the volume fraction of the oil mixture. The α-value can be calculated by using the following equation:

$$\alpha = 0.178\ (\Delta\gamma) - 0.0132 \tag{8}$$

Where $\Delta\gamma$ is the surface tension difference between ethyl acetate and oil respectively.

The surface tensions of pure oil and ethyl acetate are 30.00 mN/m [41] and 6.80 mN/m [42], respectively The $\Delta\gamma$ resulting from mixing these two substances is 23.20 mN/m, which is estimated to be γ_A. From Equation 8, α was found to be 4.10 mN/m however the obtained value of γ_{AB} from Equation 7 was 7.94 mN/m.

It is important to bear in mind that the interfacial tension of the oil/water system could be altered when another organic liquid is added to the oil phase, resulting in a compositional change at the interface and hence changes in the cohesive and adhesive forces [40]. On the basis of the above result, it is useful to rationalize whether or not the oil droplet entrapment/engulfing inside the polymer shell can occur. To be effective, this should normally occur before the solvent diffusion step, where the polymer forming the shell is in a liquid state. To achieve this goal, the oil droplet inclusion within the polymer shell has been predicted using the interfacial tension between the three phases, A, B, and C, and the spreading coefficients as defined [43, 44]:

$$S_A = \gamma_{BC} - (\gamma_{AB} + \gamma_{AC}) \tag{9}$$

$$S_B = \gamma_{CA} - (\gamma_{BC} + \gamma_{BA}) \tag{10}$$

$$S_C = \gamma_{AB} - (\gamma_{CB} + \gamma_{CA}) \tag{11}$$

Where γ is the interfacial tension (A, B, and C refer to the three phases) Based on these conventions, complete engulfing of phase A by phase C will occur if only if $S_A < 0$, $S_B < 0$ and $S_C > 0$. In this study, oil-ethyl acetate mixture is the phase A, water is phase B, and the polymeric phase (PLA) is phase C. The calculation of the spreading coefficients of a specific phase A/phase C system allows predicting the possible formation of oily core NCs.

The corresponding interfacial tension values were obtained as follows: γ_{BC} was obtained from the literature (γ_{BC} = +6.83 mN/m [45], γ_{AC} = +0.06 mN/m was obtained from Table S4 in [46, 47], and γ_{AB} calculated from Equation 7 was 7.94 mN/m, which is comparable to the literature value of 8.42 mN/m [45] The positive sign of the water-PLA interfacial tension (γ_{BC} = +6.83 mN/m) implies that it tries to reduce its energy by reducing its surface area, and therefore, a spherical shape might be maintained. Based on these data, the obtained spreading coefficients were S_A = -1.17 mN/m (< 0), S_B = -14.71 mN/m (< 0), and S_C = +1.05 mN/m (> 0). Thermodynamically, the driving forces allowed the formation of a PLA layer between the water and oily phase, thus engulfing the oily phase. These data fundamentally explain why the oily core was surrounded by a polymer layer as visually evidenced by the TEM analysis.

In Vitro Evaluation of Doc-loaded NC Cytotoxicity

Figure 7 shows the cytotoxicity from the LDH assay of native Doc and Doc-loaded NCs at three tested different concentrations. The cell culture medium used as a negative control was not cytotoxic. At lower concentrations, the native drug appeared more bioactive perhaps due to rapid diffusion and higher level of interaction with the cells. However, at the highest concentration (5 μM), the drug-loaded NCs› cytotoxicity was significantly higher than that of the native Doc. This data clearly suggests that the oily core NC formulation (as a reservoir system for the drug) could enhance the biological responses of the native Doc with higher drug payload. This might be due to the sustained release properties of the NCs and the enhanced drug internalization by SUM 225 cells in the nanocapsule form. This observation was consistent with a previous report [48] related to a nanoparticulate form of another anticancer drug (doxorubicin) tested on a different breast cancer cell line (MCF7).

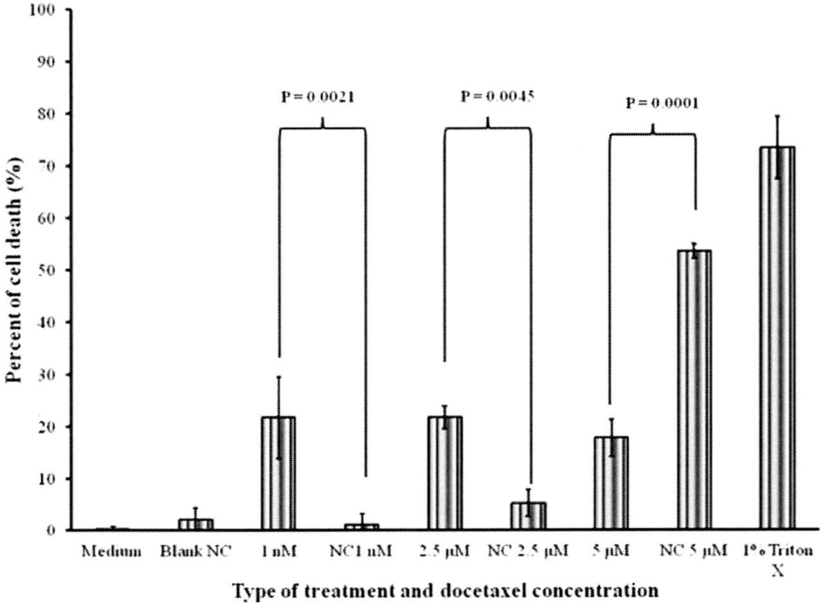

Figure 7: Percent SUM 225 cell death by LDH assay after different treatments for 24 h. See text for details. Pairwise comparison: native drug versus drug loaded NCs at the same concentration. Data are expressed as mean percent ± SD from six measurements (n = 6).

CONCLUSIONS

In this work, we developed PLA-based Doc-loaded oily NCs by solvent diffusion method. The results confirmed the formation of Doc-loaded oily core polyester-shell NCs in the nanosize range (115 to 582 nm) and high EE% (65% to 93%). Most of the NCs were monodisperse in size (PDI = 0.005) with smooth surface. The PXRD data suggested that Doc was dissolved in the NCs. The Doc release data over 144 h fitted well with the Higuchi model ($R^2 > 0.93$), indicating that the drug mainly diffused out of the NCs in this timeframe. Consistent with the analysis of the spreading coefficients and visual evidence from EM, the NCs' oily core was indeed formed. The results from the cytotoxicity study suggested that at a high concentration (5 µM), the enhanced toxicity of the encapsulated drug on the examined cancerous cell line might be due to both the particle›s uptake by the SUM 225 cells and the sustained drug release profile from the NCs. In a future work, we plan to investigate the cancer cell targeting capability of pegylated Doc-loaded NCs conjugated to specific ligands for drug delivery applications.

AUTHORS' CONTRIBUTIONS

IY designed the present work and carried out most of the experimental work (including the preparation and characterizations of the sample and the analysis of experimental data). XY participated in the analysis of the particle size, morphology, and PXRD spectra. JBM carried out the PXRD analysis. BBCY participated in the design of the study and coordination of the work as lead investigator. All authors contributed to the interpretation of the results and the drafting of the manuscript, and they read and approved the final version.

ACKNOWLEDGEMENTS

The authors are thankful to Gattefosse Corporation for providing a gift sample of Labrafac CC, to Dr. Elizabet Kostoryz (Division of Pharmacology, University of Missouri-Kansas City) for assistance with the DLS measurement, and to Randy Tindall (Electron Microscopy Center, University of Missouri-Columbia) for the electron microscopy studies. We are also grateful to Fariba Behbod, PharmD, PhD, Assistant Professor, Pathology and Laboratory Medicine, University of Kansas Medical Center for kindly providing the SUM 225 cell line.

REFERENCES

1. Herbst RS, Khuri FR: Mode of action of docetaxel - a basis for combination with novel anticancer agents. *Cancer Treat Rev* 2003, 29:407-415.
2. Aapro M, Bruno R: Early clinical studies with docetaxel. Docetaxel investigators group. *Eur J Cancer* 1995, 31A:S7-10.

3. Capri G, Tarenzi E, Fulfaro F, Gianni L: The role of taxanes in the treatment of breast cancer. *Semin Oncol* 1996, 23:68-75.
4. Rowinsky EK: The development and clinical utility of the taxane class of antimicrotubule chemotherapy agents. *Annu Rev Med* 1997, 48:353-374.
5. Pazdur R, Newman RA, Newman BM, Fuentes A, Benvenuto J, Bready B, Moore D Jr, Jaiyesimi I, Vreeland F, Bayssas MM, Raber MN: Phase I trial of Taxotere: five-day schedule. *J Natl Cancer Inst* 1992, 84:1781-1788.
6. Tomiak E, Piccart MJ, Kerger J, Lips S, Awada A, de Valeriola D, Ravoet C, Lossignol D, Sculier JP, Auzannet V, Le Bail N, Bayssas M, Klastersky J: Phase I study of docetaxel administered as a 1-hour intravenous infusion on a weekly basis. *J Clin Oncol* 1994, 12:1458-1467.
7. Bissery MC, Nohynek G, Sanderink GJ, Lavelle F: Docetaxel (Taxotere): a review of preclinical and clinical experience. Part I: preclinical experience. *Anticancer Drugs* 1995, 6:339-355. 363-338
8. Liang G, Jia-Bi Z, Fei X, and Bin N: Preparation, characterization and pharmacokinetics of N-palmitoyl chitosan anchored docetaxel liposomes. *J Pharm Pharmacol* 2007, 59:661-667.
9. Immordino ML, Brusa P, Arpicco S, Stella B, Dosio F, Cattel L: Preparation, characterization, cytotoxicity and pharmacokinetics of liposomes containing docetaxel. *J Control Release* 2003, 91:417-429.
10. Wang QW, Lu HL, Song CC, Liu H, Xu CG: Radiosensitivity of human colon cancer cell enhanced by immunoliposomal docetaxel. *World J Gastroenterol* 2005, 11:4003-4007.
11. Jakate AS, Einhaus CM, DeAnglis AP, Retzinger GS, and Desai PB: Preparation, characterization, and preliminary application of fibrinogen-coated olive oil droplets for the targeted delivery of docetaxel to solid malignancies. *Cancer Res* 2003, 63:7314-7320.
12. Einhaus CM, Retzinger AC, Perrotta AO, Dentler MD, Jakate AS, Desai PB, Retzinger GS:Fibrinogen-coated droplets of olive oil for delivery of docetaxel to a fibrin(ogen)-rich ascites form of a murine mammary tumor. *Clin Cancer Res* 2004, 10:7001-7010.

13. Farokhzad OC, Cheng J, Teply BA, Sherifi I, Jon S, Kantoff PW, Richie JP, and Langer R: Targeted nanoparticle-aptamer bioconjugates for cancer chemotherapy in vivo. *Proc Natl Acad Sci USA* 2006, 103:6315-6320.

14. Grosse PY, Bressolle F, Pinguet F: In vitro modulation of doxorubicin and docetaxel antitumoral activity by methyl-beta-cyclodextrin. *Eur J Cancer* 1998, 34:168-174.

15. Heurtault B, Saulnier P, Pech B, Venier-Julienne M-C, Proust J-E, Phan-Tan-Luu R, Benoît J-P:The influence of lipid nanocapsule composition on their size distribution. *Eur J Pharm Sci* 2003, 18:55-61.

16. Kennedy L, Bear a, Young J, Lewinski N, Kim J, Foster a, Drezek R: T cells enhance gold nanoparticle delivery to tumors in vivo. *Nanoscale Res Lett* 2011, 6:283.

17. Couvreur P, Dubernet C, Puisieux F: Controlled drug delivery with nanoparticles: current possibilities and future trends. *Eur J Pharm Biopharm* 1995, 41:2-13.

18. Wasan KM: The biological functions of lipid excipients and the implications for pharmaceutical products development. *J Pharm Sci* 2009, 98:379-382.

19. Quintanar-Guerrero D, Ganem-Quintanar A, Allémann E, Fessi H, Doelker E: Influence of the stabilizer coating layer on the purification and freeze-drying of poly (D, L-lactic acid) nanoparticles prepared by an emulsion-diffusion technique. *J Microencapsul* 1998, 15:107-119.

20. Hans ML, Lowman AM: Biodegradable nanoparticles for drug delivery and targeting. *Cur Opin in Sol State Mater Sci* 2002, 6:319-327.

21. Wilkes GM, Barton-Burke M: *Oncology Nursing Drug Handbook*. Sudbury, MA: Jones and Bartlett; 2006.

22. Cervin C, Tinzl M, Johnsson M, Abrahamsson P-A, Tiberg F, Dizeyi N: Properties and effects of a novel liquid crystal nanoparticle formulation of docetaxel in a prostate cancer mouse model. *Eur J Pharm Sci* 2010, 41:369-375.

23. Alukda D, Sturgis T, Youan B-BC: Formulation of tenofovir-loaded functionalized solid lipid nanoparticles intended for HIV prevention. *J Pharm Sci* 2011, 100:3345-3356.

24. Higuchi T: Rate of release of medicaments from ointment bases containing drugs in suspension. *J Pharm Sci* 1961, 50:874-875.
25. Hansen CM: Polymer science applied to biological problems: prediction of cytotoxic drug interactions with DNA. *Eur Polym J* 2008, 44:2741-2748.
26. Hildebrand JH, Scott RL: *The Solubility of Nonelectrolytes*. 3rd edition. New York: Reinhold; 1950.
27. Lawson DD, Ingham JD: Estimation of solubility parameters from refractive index data. *Nature* 1969, 223:614-615.
28. Hansen CM: The three dimensional solubility parameter - key to paint component affinities I. - solvents, plasticizers, polymers, and resins. *J Paint Techn* 1967, 39:104-117.
29. Latere Dwan'Isa J, Rouxhet L, Préat V, Brewster M, Ariën A: Prediction of drug solubility in amphiphilic di-block copolymer micelles: the role of polymer-drug compatibility. *Pharmazie* 2007, 62:499-504.
30. Murakami H, Kawashima Y, Niwa T, Hino T, Takeuchi H, Kobayashi M: Influence of the degrees of hydrolyzation and polymerization of poly (vinylalcohol) on the preparation and properties of poly(dl-lactide-co-glycolide) nanoparticle. *Inter J Pharm* 1997, 149:43-49.
31. Govender T, Riley T, Ehtezazi T, Garnett MC, Stolnik S, Illum L, and Davis SS: Defining the drug incorporation properties of PLA-PEG nanoparticles. *Inter J Pharm* 2000, 199:95-110.
32. Hackley V, Ferraris C: The use of nomenclature in dispersion science and technology. *NIST Recommended Practice Guide: SP* 2001.
33. Mosqueira VCF, Legrand P, Pinto-Alphandary H, Puisieux F, Barratt G: Poly (D, L-lactide) nanocapsules prepared by a solvent displacement process: influence of the composition on physicochemical and structural properties. *J Pharm Sci* 2000, 89:614-626.
34. Chen J, Qiu L, Hu M, Jin Y, Han J: Preparation, characterization and in vitro evaluation of solid dispersions containing docetaxel. *Drug Dev Ind Pharm* 2008, 34(6):588-594.
35. Zaske L, Perrin MA, Leveiller F: Docetaxel: solid state characterization by x-ray powder diffraction and thermogravimetry. *J Phys IVFrance* 2001, 11:Pr10-221-Pr10-226.

36. Abdelwahed W, Degobert G, Fessi H: Investigation of nanocapsules stabilization by amorphous excipients during freeze-drying and storage. *Eur J Pharm Biopharm* 2006, 63:87-94.
37. Sanna V, Roggio A, Posadino A, Cossu A, Marceddu S, Mariani A, Alzari V, Uzzau S, Pintus G, Sechi M: Novel docetaxel-loaded nanoparticles based on poly(lactide-co-caprolactone) and poly(lactide-co-glycolide-co-caprolactone) for prostate cancer treatment: formulation, characterization, and cytotoxicity studies. *Nanoscale Res Lett* 2011, 6:260.
38. Higuchi T: Mechanism of sustained action medication. Theoretical analysis of rate of release of solid drugs dispersed in solid matrices. *J Pharm Sci* 1963, 52:1145-1149.
39. Halder J, Landen CN, Lutgendorf SK, Li Y, Jennings NB, Fan D, Nelkin GM, Schmandt R, Schaller MD, Sood AK: Focal adhesion kinase silencing augments docetaxel-mediated apoptosis in ovarian cancer cells. *Clin Cancer Res* 2005, 11:8829-8836.
40. Kim H, Burgess DJ: Prediction of interfacial tension between oil mixtures and water. *J Colloid Interface Sci* 2001, 241:509-513.
41. Adamson A, Gast a: *Physical Chemistry of Surfaces*. Hoboken: Wiley; 1997.
42. Kobayashi I, Mukataka S, and Nakajima M: Effects of type and physical properties of oil phase on oil-in-water emulsion droplet formation in straight-through microchannel emulsification, experimental and CFD studies. *Langmuir* 2005, 21:5722-5730.
43. Pekarek KJ, Jacob JS, Mathiowitz E: Double-walled polymer microspheres for controled drug release. *Nature* 1994, 367:258-260.
44. Torza S, Mason SG: Three-phase interactions in shear and electrical fields. *J Colloid Interface Sci* 1970, 33:67-83.
45. Howard JR, Hadgraft J: The clearance of oily vehicles following intramuscular and subcutaneous injections in rabbits. *Int J Pharm* 1983, 16:31-39.
46. Biresaw G, Carriere CJ: Interfacial tension of poly (lactic acid)/polystyrene blends. *J Polym Sci B Polym Phys* 2002, 40:2248-2258.

47. Kirchner B: *Ionic Liquids*. *Volume 290*. Berlin: Springer; 2010.

48. Li PY, Lai PS, Hung WC, Syu WJ: Poly (l-lactide)-vitamin E TPGS nanoparticles enhanced the cytotoxicity of Doxorubicin in drug-resistant MCF-7 breast cancer cells. *Biomacromolecules* 2010, 11:2576-2582.

Chapter 8

Assessment of Heavy Metal (Cu, Ni, Fe, Co, Mn, Cr, Zn) Pollution in Effluent Dominated Rivulet Water and Their Effect on Glycogen Metabolism and Histology of Mastacembelus Armatus

Mehjbeen Javed and Nazura Usmani

Aquatic Toxicology Research Laboratory, Department of Zoology, Aligarh Muslim University, Aligarh 202002, India

ABSTRACT

The present study was conducted to examine the contamination of rivulet situated at Kasimpur, Aligarh (27.218° N; 79.378° E). It receives the wastewater of Harduaganj Thermal Power Plant (HTPS) containing fly ash and heavy metals. Among the heavy metals estimated in the rivulet water, Fe (8.71 mgL^{-1}) was present in the highest concentration followed by Cu (0.86 mgL^{-1}), Zn (0.30 mgL^{-1}) Mn (0.21 mgL^{-1}), Ni (0.12 mgL^{-1}), Co (0.11 mgL^{-1}) and Cr (0.10 mgL^{-1}). The values for the heavy metals such as Fe, Ni and Mn were beyond the limits set by UNEPGEMS. Bioaccumulation of these heavy metals was detected in tissues such as gills, liver, kidney, muscle and integument of the fish *Mastacembelus armatus*. Accumulation of Fe (213.29 – 2601.49 $mgkg^{-1}$.dw) was highest in all the organs. Liver was the most influenced organ and integument had the least metal load. The accumulation of Fe, Zn, Cu and Mn, observed in the tissues were above the values recommended by FAO/WHO. Biochemical estimation related to blood glucose, liver and muscle glycogen conducted showed significant ($p < 0.01$) elevation in blood glucose content over control (17.73%), whereas liver glycogen dropped significantly ($p < 0.01$) over control (−89.83%), and similarly muscle glycogen also decreased significantly ($p < 0.05$) over control (−71.95%), suggesting enhanced glycolytic capacity to fuel hepatic metabolism. Histopathological alterations were also observed in selected organs (gills, liver and kidney) of *Mastacembelus armatus*.

INTRODUCTION

Sewage and industrial disposal has greatly increased the addition of heavy metals in the aquatic ecosystems. In Aligarh as well there are very few studies to focus on this aspect. It influences the productivity and health status of water bodies as abnormal changes in physicochemical conditions and other quality parameters have their impact on diversity. Harduaganj Reservoir (27.218° N and 79.378° E) at Kasimpur, (during late 1990's) Aligarh was quite productive and healthy (Figure 1). Water filled area being 13.5 ha. Now this reservoir is damaged completely because of the discharge of effluents from Harduaganj Thermal Power Station. This has changed the condition of reservoir drastically (Figure 2). We are being deprived of the good resource of water and fish food. It appears more of a waste land. Besides the damage of aquatic ecosystem the fly ash from the power plant has also destroys the terrestrial ecosystem. The area was previously occupied by the lush green crops of mustard, wheat and rice. With time the greenery has been replaced by waste land having ash deposited over it. This all happens due to unsafe disposal of ash from the Power station damaging the soil making it unfit for agriculture. The pipelines from the Power Plant are now used to discharge their wastewaters in one of the adjacent rivulet flowing through the area (Figure 3). Several species of fishes are thriving in such water bodies hence their conservation is essential. The Thermal Power Plant effluent contains gases, fly ash and traces of heavy metals which can endanger the inhabitants. The rivulet water has also been rendered unsafe for fishes as well as for domestic consumption, irrigation, and other needs particularly consumption by cattle and other domestic animals. This attracted a lot of attention as another aquatic ecosystem may get damaged and have the same fate as the reservoir. Therefore the construction of power plants in the areas near crop fields and water bodies should not be prevalent. This will check its impact on both agriculture and aquaculture especially from the capture fisheries point of view and small water bodies usually meet the need of local population.

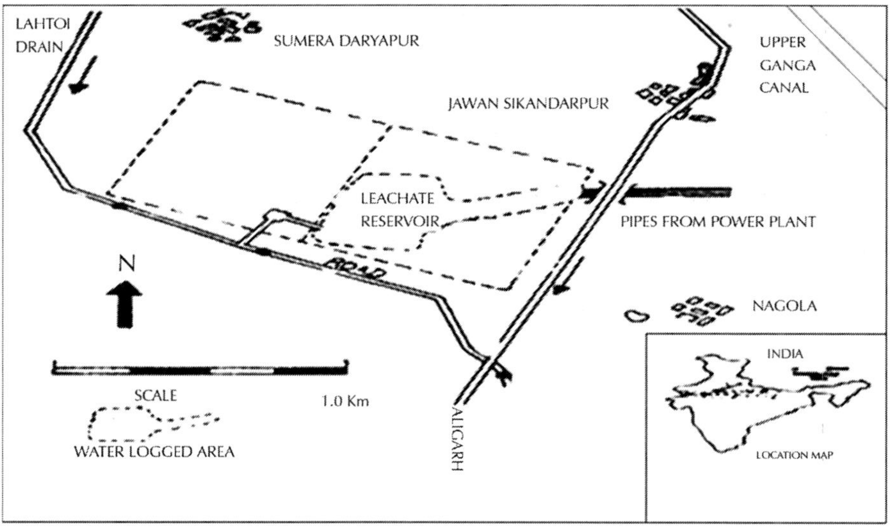

Figure 1: Map showing the location of Harduaganj Reservoir (27.218° N and 79.378° E) which receives the effluents from the Power Plant via pipes.

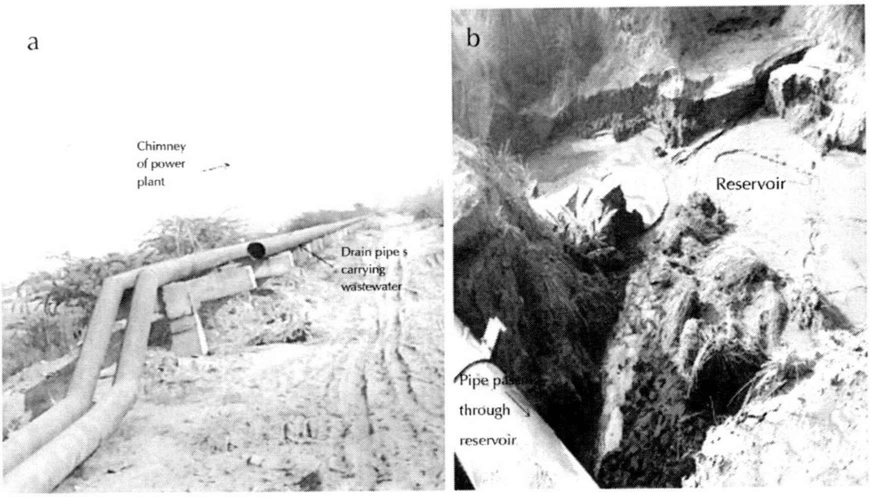

Figure 2: a) Harduaganj Thermal Power Plant draining wastewater into the reservoir b) Present Scenario: Degraded condition of reservoir.

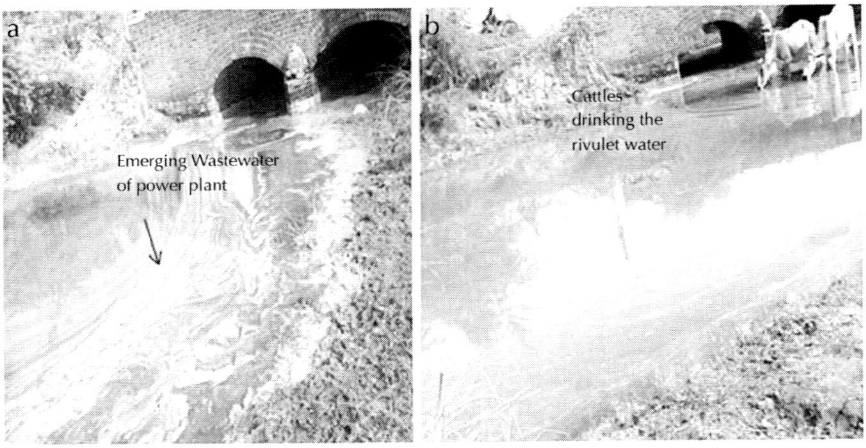

Figure 3: a) Show wastewater emerging from Power Plant polluting the rivulet b) Rivulet water used for different Purposes.

The aim of this study was to assess the abnormal presence of heavy metals in rivulet water located near Harduaganj Reservoir, their bioaccumulation in different tissues of fish, depletion/increase in carbohydrate levels in blood, muscle and liver and histopathology in organs of the fish. For this purpose, the bioaccumulation of Cu, Ni, Fe, Co, Mn, Cr and Zn were investigated in the fish tissues like gills, liver, kidney, muscle and integument. The histological analysis of gills, kidney and liver was performed in order to confirm the structural damage. Blood glucose, liver and muscle glycogen were also determined as the stress indicators of the fish health. *Mastacembelus armatus* being more prevalent and relished, also of its hardy and tolerant nature and adaptability to various conditions and habitats, was chosen as the model for study.

There is no data on the effect of heavy metals on *Mastacembelus armatus* in general and bioaccumulation, biochemical aspects and histopathology in particular. Data generated from this study will serve as an important reference in the implementation of management strategies against heavy metal pollution, their accumulation in freshwater ecosystems and conservation of aquatic life and water bodies. This study involve the use of already exposed fish from polluted waters which restricts the misuse and unnecessary killing of healthy fishes and also avoid the ethical issues regarding the use of animals in the experiments.

MATERIALS AND METHODS

Sample Collection

Water was collected in a pre-cleaned and acidified glass bottles and preserved by acidifying with 6 N HNO_3 (pH maintained to about 2.0) to estimate the presence of heavy metals in the rivulet.

On spot fixation of water was carried out to measure the dissolved oxygen (D.O). Total solids (T.S), total dissolved solids (T.D.S) and total suspended solids (T.S.S) were determined using standard techniques (APHA 2005). The temperature and pH were recorded on the spot using thermometer (Deluxe, 6) and pH strips (S.D Fine chemicals, 0–0.1).

Samples of *Mastacembelus armatus* (n=25) were collected with the help of professional local fishermen. Live specimens of *M. armatus* were transferred to water buckets and brought to the laboratory for further analysis. Sampled fishes were immediately killed, measured and then utilized for analysis of bioaccumulation of heavy metals, histopathology and biochemical parameters.

Estimation of Heavy Metals in Water Samples and Fish Organs/Tissues

Heavy metals (Cu, Ni, Fe, Co, Mn, Cr and Zn) were estimated in rivulet water and fish organs/tissues using Atomic Absorption Spectrometer (Perkin Elmer, Analyst a 800) as per the standard protocols of APHA (2005).

Biochemical Analysis

Blood Glucose Estimation

Fresh fishes were used for blood collection through cardiac puncture and the blood was placed into the sodium fluoride tubes. Blood samples taken were centrifuged at 3500 rpm for 10 min to obtain serum. The glucose levels in the serum were analyzed using the reagent Eco-Pak Glucose (Accurex Biomedical Pvt. Ltd, India). The glucose levels in samples were measured spectrophotometrically (UV–VIS Systronics, 118) against blank at 505 nm.

Liver and Muscle Glycogen Estimation

Fresh fishes were dissected to remove the liver and muscles to estimate the glycogen. The glycogen levels in liver and muscle were measured by Anthrone reagent according to the protocols of Carroll et al. (1956).

Tissue Preparation for Histological Examination

Fresh fishes were dissected to remove the organs (gills, kidney and liver) which were then fixed in bouin's fluid. Tissues were processed according to the protocols of Humason (1979). All sections were examined and photographed using Nikon Eclipse 80*i* microscope.

Metal Pollution Index

The metal Pollution index (MPI) is used to compare the total metals accumulation level in various tissues of the fish when they are beyond five in number. The values were calculated using the equation:

$$MPI = (Cf_1 x Cf_2 x\_\_\_\_\_ Cfn)^{1/n}$$

Where, *Cfn* is the contents for the metal *n* in the sample (User et al. 1997)

Bioaccumulation Factor

The bioaccumulation of the heavy metals (HM) in fish tissues were quantified using bioaccumulation factor (BAF). BAF is the ratio of the concentration of a specific heavy metal in the tissue of the organism, to the concentration that heavy metal in the water.

$$BAF = \frac{\text{Concentration of HM in dry fish tissue } (\text{mg Kg}^{-1})}{\text{Concentration of HM in rivulet water } (\text{mg L}^{-1})}$$

Statistical Analysis

All values are given as Mean ± SD. Statistical differences among the means of heavy metal accumulation in fish tissues were calculated using ANOVA and Duncan's Multiple Range Test (Duncan1955). Blood glucose and liver and muscle glycogen estimations were statistically analyzed using student's t-test (2 tailed) with the help of SPSS 17.

RESULTS

The average length of fishes measured was 14.20 ± 1.5 cm and average weight was 35 ± 0.42 g.

Physicochemical Parameters and Heavy Metals in Rivulet Water Samples

Physicochemical characteristics of rivulet water (Temperature = 27.60°C; pH = 6.9; Dissolved oxygen = 6.9 mgL^{-1}; Total solids = 652 mgL^{-1}; Total dissolved solids = 407 mgL^{-1}; Total suspended solids = 245 mgL^{-1}) were comparable with ideal water quality. However the heavy metal content in rivulet water were in the order Fe (8.71 mgL^{-1}) > Cu (0.86 mgL^{-1}) > Zn (0.3 mgL^{-1}) > Mn (0.21 mgL^{-1}) > Ni (0.12 mgL^{-1}) > Co (0.11 mgL^{-1}) > Cr (0.10 mgL^{-1}) (Table 1) where it was found that Fe, Ni, Mn and Cr exceeded the recommended values set by UNEPGEMS (2006) but Cu and Zn were within the tolerable range.

Table 1: Heavy metal content in rivulet water compared with Quality Guidelines and Standards by International Organization or country

Heavy metals	WHO (guidelines)	USA (standards)	*Present study
Cu	2	1.3	0.86±0.01
Ni	0.02	-	0.12±0.02
Fe	-	0.3	8.71±2.88
Co	-	-	0.11±0.02
Mn	0.5	0.05	0.21±.10
Cr	0.05	0.1	0.1±0.02
Zn	3	5	0.3±0.03

All values are in mgL^{-1}.

*Values of heavy metal content in the present study are given as Mean±SD, (n=4×3), samples collected from 4 different zones of rivulet and were analyzed in triplicates.

Adapted for Water Quality for Ecosystem and Human Health, 2006 (prepared and published by the United Nations Environment Programme.

Global Environment Monitoring System (GEMS)/ Water Programme).

Blank cells indicate that no, citable information was available.

Javed and Usmani

Javed and Usmani *SpringerPlus* 2013 **2**:390, doi:10.1186/2193-1801-2-390

Heavy Metals in Fish Tissue Samples

Table 2 (superscripts) shows the accumulation of heavy metals in organs of *Mastacembelus armatus*, and subscript shows the accumulation of a heavy metal in various organs of the fish.

Table 2: Accumulation of heavy metals in the organs of *Mastacembelus armatus*

Heavy metals	Gills	Liver	Kidney	Muscle	Integument
Cu	c199.88b±0.20	d271.67a±1.15	c175.89c±0.19	d41.36d±0.54	d36.27e±0.14
Ni	c200.00b±1.73	c449.96a±0.06	d149.33c±0.50	c58.98d±0.09	c45.062e±0.02
Fe	a799.66b±0.41	a2601.49a±0.50	a649.76c±0.68	a213.29e±0.31	a313.36d±0.31
Co	ND	f25.66a±0.57	ND	e9.06b±0.06	f9.06b±0.05
Mn	d25.36b±0.62	e49.96a±0.05	ND	e9.03d±0.06	e13.62c±0.00
Zn	b549.33b±0.57	b1741.95a±0.06	b351.28c±0.48	b186.19d±0.18	b168.11e±0.10

Values are Mean±S.D, (n=15), ND= not detected.

Duncans Multiple Range test was used to test the significance among the means.

Means with similar letters (a, b, c, d,e, f) in a column and row are statistically similar at $p < 0.01$.

**Superscripts indicate accumulation of different heavy metals (mgKg^{-1}.dry weight) in particular organs.*

***Subscripts indicate accumulation of a heavy metal (mgKg^{-1}.dry weight) in various organs.*

Javed and Usmani

Javed and Usmani SpringerPlus 2013 ***2****:390, doi:10.1186/2193-1801-2-390*

Fe

The concentration of Fe in different organs ranged from 213.29 to 2601.49 mg kg^{-1}.dw and it was highest (2601.49 mg kg^{-1}.dw) in liver and least (213.29 mg kg^{-1}.dw) was recorded in muscles. The order of accumulation was liver > gills > kidney > integument > muscle.

Zn

The Zn concentration fluctuates between 168.11 to 1741.95 mg kg^{-1}. dw and its pattern of accumulation was liver > gills > kidney > muscle > integument.

Ni

Similarly, Ni concentration was highest (449.96 $mgkg^{-1}$.dw) in liver and least (45.06 $mgkg^{-1}$.dw) in integument and the sequence of their presence in organs/tissues were liver > gills > kidney > muscle > integument.

Cu

Cu ranged from 36.27 to 271.67 mg kg^{-1}.dw in different organs/tissues. Its accumulation was highest in liver and least in integument. The pattern of accumulation was liver > gills > kidney > muscle > integument.

Mn

Mn concentration was highest (49.96 $mgKg^{-1}$.dw) in liver and least (9.03 $mgKg^{-1}$.dw) in muscle. The pattern of accumulation was in the order liver > gills > integument > muscle. Mn was not detected in the kidney of *Mastacembelus armatus*.

Co

Co concentration was maximum (25.66 mg kg^{-1}.dw) in liver followed by muscle (9.06 mg kg^{-1}.dw) and integument (9.06 mg kg^{-1}.dw). Values obtained for Co in muscle and integument was statistically insignificant. It was not detected in gills and kidney.

Fe accumulation was highest amongst all organs.

All the values recorded for heavy metals accumulation in different organs/tissues were statistically significant ($p > 0.01$) except for Ni and Cu in gills and Co and Mn in muscles.

According to the metal pollution index calculated for the sampled organs/tissues (Table 3) it was observed that liver (298.68) of the fish most influenced and had highest metal load followed by kidney (278.119), gills (213.63) and integument (46.86). Muscles (44.64) was least influenced by heavy metals. This clearly indicates that each tissue have different capacity of accumulation.

Table 3: Metal Pollution index value of total metal accumulation in organs/tissues of *Mastacembelus armatus*

Organs/tissues	Metal Pollution Index (MPI)
Liver	298.68
Kidney	278.19
Gills	213.63
Integument	46.86
Muscle	44.64

Javed and Usmani

Javed and Usmani *SpringerPlus* 2013 **2**:390, doi:10.1186/2193-1801-2-390

Since muscle and integument are edible part and therefore their quality needs to be monitored prior to consumption. The accumulation of heavy metals needed to be compared with recommended levels (Table 4). It was observed that except for Ni the concentrations of Cu, Fe, Mn and Zn are beyond the safe limits.

Table 4: Comparative account of heavy metal concentrations in edible part with standard guidelines

Heavy metals	*Average concentration	Recommended levels (ppm)
Cu	38.815	30 (FAO/WHO 1983)
Ni	52.04	70–80 (USFDA 1993b)
Fe	263.325	100 (FAO/WHO 1989)
Co	9.06	-
Mn	11.325	1.0 (FAO/WHO 1989)
Zn	177.15	100 (FAO/WHO 1989)

*At times along with fish muscle, integument is also consumed therefore average of both is taken.

Blank cells indicate that no information was available.

Javed and Usmani

Javed and Usmani *SpringerPlus* 2013 **2**:390, doi:10.1186/2193-1801-2-390

Table 5 revealed the calculated bioaccumulation factor for different heavy metals in fish tissues. It was found that the concentration of the heavy metals in different tissues of the fish were several folds higher than their concentrations in water. Liver showed high bioaccumulation factor (BAF) for all the heavy metals. In gills the highest BAF was of Zn and in kidney it was Ni. Similarly in muscle the highest BAF was recorded for Zn and least for Fe. In integument highest and least values of BAF were calculated for Zn and Fe respectively.

Table 5: Bioaccumulation factors (BAF)* of heavy metals in the different tissues of*Mastacembelus armatus*

Heavy metals	Gills	Liver	Kidney	Muscle	Integument
Cu	232.41	315.89	204.52	48.09	42.17
Ni	1666.66	3749.66	1244.41	491.50	375.51
Fe	91.80	298.67	74.59	24.48	35.97
Co	ND	233.27	ND	82.36	82.36
Mn	120.76	237.90	ND	43.00	64.85
Zn	1831.10	5806.50	1170.93	620.63	560.36

$$^{*}BAF = \frac{\text{Concentration of HM in dry fish tissue } (mg\,kg^{-1})}{\text{Concentration of HM in rivulet water } (mgL^{-1})}$$

Javed and Usmani

Javed and Usmani *SpringerPlus* 2013 **2**:390, doi:10.1186/2193-1801-2-390

Biochemical Analysis

It is evident from Table 6 that blood glucose showed significant ($p < 0.01$) elevation (+17.73%) over control. Whereas liver glycogen content depleted (−89.83%) significantly ($p < 0.01$) when compared to control. Similarly, muscle glycogen decreased (−71.95%) significantly ($p < 0.05$) over control.

Table 6: Impact of Heavy metal concentrations on glucose and glycogen content of *Mastacembelus armatus*

Tissues	Control	Exposed	Percent change over control
	Glucose/ Glycogen content	Glucose/ Glycogen content	
Blood (glucose)	1.14±0.01**	21.36±0.049**	+17.73%
Liver (glycogen)	2.36±0.01**	0.24±0.111**	−89.83%
Muscle (glycogen)	0.82±0.01*	0.23± 0.032*	−71.95%

Values are Mean±S.D (n=8); Blood glucose is given in mg%; Glycogen values are given in mg/g; Student's t- test was used to test the significance **Significant at $p < 0.01$ and *Significant at $p < 0.05$; +/− indicates increase or decrease over control.

Javed and Usmani

Javed and Usmani *SpringerPlus* 2013 **2**:390, doi:10.1186/2193-1801-2-390

Histopathological Observations

Histopathological examinations confirmed the deformities in gills, kidney and liver of the fish and results are shown in Figure 4, 5 and 6. Gill of control fish Figure 4 (a) showed normal structure while exposed fish sections Figure (b,c,d) showed severe lamellar fusion, hyperplasia, hypertrophy and epithelial lifting, swelling and deformed lamella, in some parts sloughing off and curving of lamellae was also observed. Figure 5 (a) Showed liver of control *Mastacembelus armatus* which had network of hepatocytes and bile duct. While Figure 5 (b and c) revealed the histological damage that occurred in liver, necrosis of parenchyma, vacuolation, congestion of blood vessel, pyknosis and infiltration of leucocytes were the major observations. Figure 6 (a) Normal kidney of fish and Figure 6 (b and c) describes the dilation and vacuolation of kidney tubules, decreased lumen of tubule due to hypertrophy, glomeruli with no bowman's space, degeneration of glomeruli and necrosis of hematopoietic tissue.

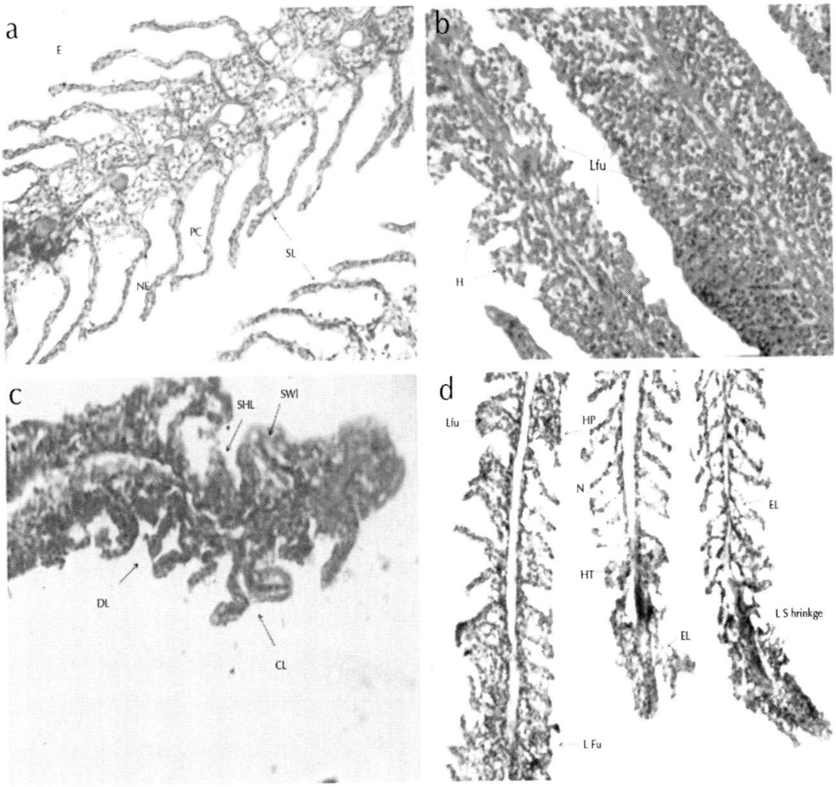

Figure 4: a) Gills of control *Mastacembelus armatus* show normal structure of lamellae, erythrocyte (E), nucleus of epithelial cell (NE), pillar cell (PC), secondary lamellae (SL) 400X; Figure b, c and d) Gills of exposed *Mastacembelus armatus* showing b) complete lamellar fusion (Lfu) coupled with hyperplasia (H) 200X; c) Shrinked lamella (SHL), swelling of lamella (SWL), deformed lamella (DL), curved lamella (CL) 400X; d) Lamellar fusion (Lfu), hyperplasia (HP), necrosis (N), Hypertrophy (HT), epithelial lifting (EL), Lamellae shrinkage (L Shrinkage), Stained with Hematoxylin and Eosin.

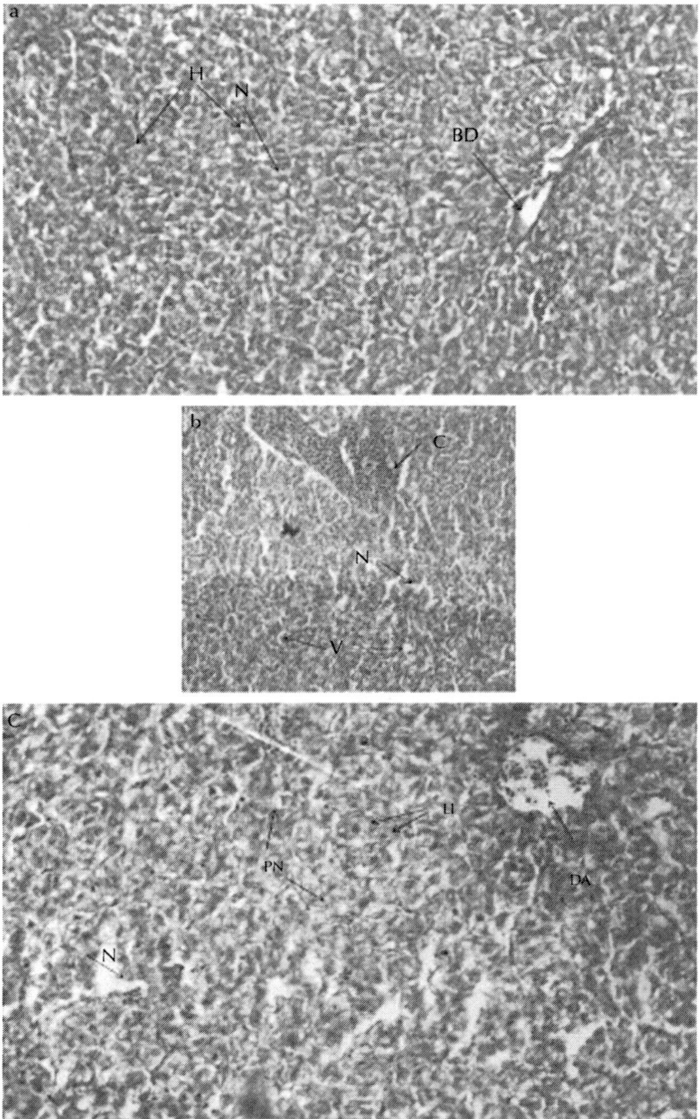

Figure 5: a) Show control liver of *Mastacembelus armatus* having hepatocytes (H) with normal nucleus (N), bile duct is also present 400X; Figure b and c) Show liver of exposed *Mastacembelus armatus*; b) Congestion of blood vessel (C), necrosis (N) and vacuolization (V) of hepatocytes 200X; c) Pyknotic nuclei (PN), infiltration of leucocytes (LI), degeneration of arteriole (DA) and necrosis of parenchyma (N) 400X. Stained with Hematoxylin and Eosin.

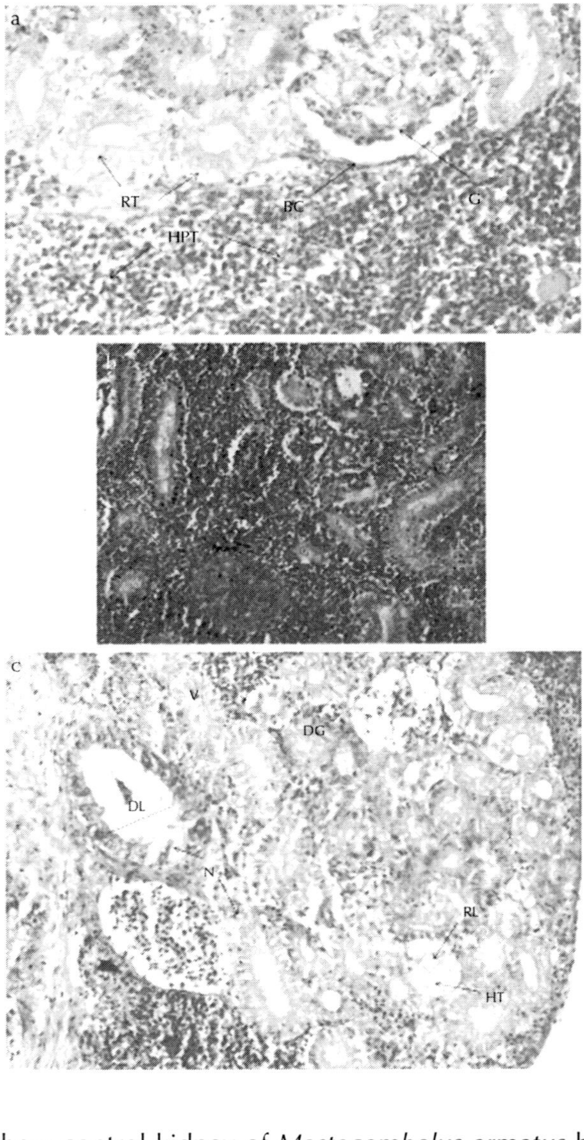

Figure 6: a) Show control kidney of *Mastacembelus armatus* having normal structure glomeruli (G) with bowman's capsule (BC), renal tubules (RT) and hematopoeitic tissues (HPT) 400X; Figure b and c) Show kidney of exposed *Mastacembelus armatus* b) Vacuolization of tubules (V), hyperplasia of glomeruli (GHY) and have no bowman's space (NBS), necrosis of renal tubule (NT) 200X; c) Lumen of tubule dilated (DL), necrosis of hematopoeitic tissue and tubule (N), hypertrophy of tubule (HT) due to which lumen is reduced (RL) 400X; Stained with hematoxylin and eosin.

DISCUSSION

The water quality of rivulet was suitable for sustenance of fishes as determined by physicochemical characteristics however among the heavy metals examined some were beyond the maximum permissible limits. The heavy metal content estimated in water were present in the order Fe > Cu > Zn >Mn> Ni > Co > Cr where it was found that Fe, Ni, Mn and Cr content exceeded the recommended guidelines set by UNEPGEMS (2006) due to which the water becomes unfit for the inhabitant fishes. There are very few studies which reported presence of heavy metals in waters of Aligarh as well. Javed and Usmani (2012a) reported that water of sewage fed aquaculture pond at Panethi, Aligarh which is used as a source of commercial fish food also contain heavy metals in the order Fe >Mn> Zn > Co > Ni > Cu = Cr. In another study the sugar mill effluent dominated river at village Satha, Aligarh also reported to contain heavy metals in the order Ni > Cr > Cu > Co (Javed and Usmani 2013b). But the metal content in these waters are lower than the present study. Similarly many water bodies which lie in vicinity of population have been polluted by effluents released by industries, factories, Power stations, domestic waste etc. which besides disturbing the quality of water also degrade the protein source in the form of fish food and limits their use (Baki et al. 2011; Abdul Qadir and Riffat Naseem 2011; Javed and Usmani 2012b 2013a; Taweel et al. 2012; Emere and Dibal 2013; Fatima and Usmani 2013). In the present study possible reason for high levels of heavy metals in rivulet water could be attributed to the fact that this Power Station may not have efficient wastewater treatment plant. Although these metals are essential in traces but their excess becomes toxic and gets accumulated in organs/tissues of fish *Mastacembelus armatus*.

Bioaccumulation of Heavy Metals in *Mastacembelus Armatus*

Fe

In the present study it was examined that Fe content was high in water and the same was observed in organs of *Mastacembelus armatus*. It had highest accumulation in liver and least was recorded in muscle. Studies reported for *Clarias gariepinus* (Osman Alaa and Werner 2009), *Tinca tinca* (Selda Tekin et al. 2005), *Labeo rohita* (Javed and Usmani 2011) also revealed the maximum accumulation of Fe in liver and least in muscle. The high Fe concentrations observed in the liver tissue of *M. armatus* could be due to iron-containing enzymes and the extensive vascular system of the liver, as the haemoglobin in the blood binds approximately three quarters of the Fe in the body (Voynar1960). As compared with the present study the values reported for Fe content in different tissues of*Channa punctatus* inhabiting sewage fed pond, Aligarh was higher and highest accumulation occurred in gills and least in integument (Javed and Usmani 2012a). Accumulation of Fe in muscle of*M. armatus* examined was beyond the permissible limits (100 ppm) set by FAO/WHO (1989).

Fe was highest accumulated metal among Cu, Ni, Co, Mn, Cr, Zn in all the organs/tissues of fish including the present study which may be due to the unique feature of iron metabolism, that there was no route for complete excretion for Fe. Consequently, little iron is lost. High doses of iron are frequently associated with gastrointestinal effects, especially constipation, but also with nausea, diarrhoea and vomiting. Chronic Fe exposure promotes gastric and esophageal ulceration in humans and animals.

Zn

Fe accumulation was followed by Zn in organs/tissues of *Mastacembelus armatus*. In the present study it was observed that Zn accumulation was highest in liver and least in integument. Highest accumulation of Zn in liver has also been reported in *Clarias gariepinus* (Coetzee et al. 2002),*Channa punctatus* (Murugan et al. 2008) and *Labeo rohita* (Javed and Usmani 2011) and least in integument of *Channa punctatus* (Javed and Usmani 2012a). The reason for high accumulation of Zn in liver could be due to metallothionein (MT). These are low molecular weight proteins and their production increases due to elevated levels of heavy metals which bind to the metal in order to detoxify them, in doing so they concentrate and regulate metals in liver. In the present study it has also been observed that Zn content was lower in muscle than liver which could be due to the transference of Zn from muscle to other fish organs, and this deloading ability of fish has been reported to be advantageous to fish consumers (Madhusudan et al. 2003; Murugan et al. 2008). Zn accumulation was beyond the permissible limits (100 ppm) set by FAO/WHO (1989) in muscle of *M. armatus*. Although Zn is an essential element as it is carefully regulated by physiological mechanisms in most organisms (Eisler 1988), it is also regarded as a potential hazard that can endanger both animal and human health. Agency for Toxic Substances and Disease Registry (ATSDR 2005), suggests that ingesting high levels of zinc for several months may cause anemia, damage the pancreas, and decrease levels of high-density lipoprotein (HDL) cholesterol.

Ni

Ni occupied the third position as far as bioaccumulation was concerned. Highest accumulation was observed in liver and least in integument. In other studies as well the liver reported to accumulate high concentrations of Ni in fishes such as *Labeo umbratus* (Coetzee et al. 2002), *Oreochromis niloticus* (Taweel et al. 2012), *Cirrhinus mrigala* (Javed 2012) and least in integument of *Clarias gariepinus*, *Channa punctatus* (Javed and Usmani 2012a). Accumulation of Ni in muscle of *M. armatus* was below the permissible limits (70–80) set by USFDA (1993b). Ni content has also been reported to be beyond the permissible limits in water of Yamuna River at Okhla barrage as well in fishes *Channa striatus* and *Heteropneustes fossilis* (Fatima and Usmani 2013). Ni exhibits similar chemical behavior as Fe and Co. Concentrations of Ni in water is likely to be of health concern in environments where pH is less than 4.5. As is the case with other essential elements Ni is also toxic to fish when present in high enough concentrations (Pickering 1974). An uptake of too large quantities of nickel results in respiratory failure, Birth defects, Asthma and chronic bronchitis, Heart disorders etc. This can also cause various kinds of cancer on different sites within the bodies of animals.

Cu

Highest accumulation of Cu also occurred in liver and least in integument. Other workers also reported the highest copper content in liver of fishes *Labeo umbratus* (Coetzee et al. 2002),*Oncorhynchus mykiss*, *Cyprinus carpio* (Boeck et al. 2004), *Wallago attu* (Yousafzai et al. 2010). The Cu content observed in the present study in muscle is similar to accumulation observed in *Clarias gariepinus* (Anim et al. 2010) and *Wallago attu* (Yousafzai et al. 2010). According to Stokes (1979), fish muscles have poor accumulative properties, even in systems containing high copper levels. The accumulation of Cu in muscle was beyond the permissible limits (30 ppm) set by FAO (1983). But in another study from sugar mill effluent dominated water body at Satha, Aligarh, the Cu content observed in edible part of *Channa punctatus* was below the permissible limits (Javed and Usmani2013b). Copper is essential for good health. However, exposure to higher doses can be harmful. Cu toxicity in natural water arising from pollutants may cause severe damage in gills and necrotic changes in the liver and kidneys. According to ATSDR (2005), long term exposure to Cu, higher than normal levels can cause nausea, vomiting, stomach cramps, or diarrhea.

Mn

The highest accumulation of Mn was observed in liver and least in muscle. However other workers investigated highest accumulation of Mn in gills and least in muscle of fishes such as *Oreochromis mossambicus* (Robinson and Oldewage 1997), *Tinca tinca* (Selda Tekin et al. 2005), *Clarias gariepinus* (Osman Alaa and Werner 2009), *Cyprinus carpio* (Jabeen and Chaudhry 2010), *Clarias anguillaris* (Nwajei et al. 2012). Accumulation observed in muscle in present study corroborates to*Oreochromis mossambicus* (Robinson and Oldewage 1997) and *Clarias gariepinus* (Coetzee et al.2002). The content of Mn examined in muscle of *M. armatus* was several folds higher than the maximum permissible limits (1.0 ppm) of Mn set by FAO/WHO (1989). High levels of Mn can cause lung, liver and vascular disturbances, declines in blood pressure, failure in development of animal foetuses and brain damage. Finally, laboratory tests with test animals have shown that severe manganese poisoning should even be able to cause tumor development in animals.

Co

Co was the least accumulated metal in tissues of *Mastacembelus armatus*. The highest accumulation occurred in liver while muscle and integument were least influenced. Other workers reported high accumulation of Co in kidney of fishes *Channa punctatus* and *Labeo rohita* but comparable levels in muscle and integument (Javed and Usmani 2011). Smith and Carson (1981) reported that Co is mostly accumulated (about 75%) in the viscera and integument of the fish and the statement agrees well with the present study. No permissible guidelines/limits have yet been established for Co. Generally, cobalt compounds that dissolve easily in water are more harmful than those that are hard to dissolve in water. Once cobalt enters the body, it is distributed into all tissues, but mainly into the liver, kidney, and bones (ATSDR 2004).

Cr

In the present study it was not detected in any of the fish organs/tissues though it was reported beyond permissible limits in rivulet water. However high concentrations had been reported in the water of sugar mill effluent dominated water body at Satha, Aligarh as well as in prevalent fish*Channa punctatus* (Javed and Usmani 2013b). Similarly Fatima and Usmani (2013) also reported Cr content in both water and fishes (*Channa striatus* and *Heteropneustes fossilis*) of Yamuna at Okhla barrage, which was beyond the permissible limits.

In this study, liver was observed to be the most influenced organ. Lower concentrations in muscle and integument possibly indicate that the fish integument is an important excretory organ for these metals and also serve as a protective barrier between outer environment and muscle. But still the accumulated amount is of concern.

Bioaccumulation Factor (BAF)

Bioaccumulation factor was used to quantify the metal concentrations in fish organs/tissues relative to the concentrations in rivulet water. In the present study the BAF values was high. It was calculated highest for liver (233.27- 5806.5) and least for integument (35.97- 560.36) and following the order, liver > gills > kidney > muscle > integument. This showed that liver had high metal load than other organs.

Effects of Heavy Metals on Glycogen Metabolism

The Thermal Power Plant effluents brought about a number of significant changes in the carbohydrate metabolism in *Mastacembelus armatus*. There was significant elevation in blood glucose over control (+17.73%) while glycogen in liver (−89.83%) and muscle (−71.95%) decline significantly. This hyperglycemic (increase in blood glucose) condition were also detected in fishes, *Heteropneustes fossilis* and *Saccobranchus fossilis*, exposed to Ni and Cr (Nath and Kumar 1988; Radhakrishaniah et al. 1992), *Labeo rohita* and *Clarias gariepinus* subjected to Cu (Van Vuren et al. 1994; James et al.1996), *Oreochromis aurues* exposed to mixture of Cu and Pb salts. During the present study increase in blood glucose content as a result of heavy metals present in power plant effluent could be attributed to the enhanced glycogenolysis, resulting in formation of glucose to meet the energy demand during stress.

In the present study liver and muscle glycogen dropped significantly under the stress of heavy metals. Srivastava and Srivastava (2008) reported that glycogen consistently decreased from 8.18 to 5.3 mg g^{-1}in *Channa punctatus* when exposed to sublethal concentrations of $ZnSO_4$. Similarly, other studies also subscribes to the above view in fishes such as *Channa punctatus* subjected to distillery effluent (Maruthi and Subba Rao 2000), *Mystus cavasius* to electroplating industrial effluent (Palanisamy et al. 2011). In this study it has been noticed that liver has highest Metal Pollution Index therefore it is assumed that high accumulation levels of heavy metals in liver impaired the activity of enzymes which contribute to glycogen synthesis, leading to decrease in glycogen content.

Histopathology

The present study revealed that Thermal Power Plant effluents induced histopathological alterations in gills, liver and kidney of *Mastacembelus armatus*. Under present investigation, it has been observed that gills of fish exhibited several histological alterations like complete fusion of lamellae coupled with hyperplasia, hypertrophy and epithelial lifting, necrotic and shrinked/curved lamellae. Similar results have been reported in fishes such as *Labeo rohita* exposed to Cr (VI) (Sesha Srinivas V and Rao BM1998), *Tilapia mossambica* to Cu, Ni, Cr (Ravanaiah and Narasimha Murthy CV 2010), *Cyprinus carpio* to Cr (Parvathi et al. 2011), *Clarias gariepinus* to sewage and domestic wastewater containing Cu, Fe, Pb, Cd and Zn (Authman et al. 2013). The alterations observed in gill of *M. armatus* prove that it was under chronic exposure and these changes could be interpreted as defense responses of the fish, as these alterations increase the distance across which the dissolved heavy metals must diffuse to reach the bloodstream. Prolonged exposure to heavy metals can lead to degeneration of the epithelium. It can therefore be argued that gill epithelium was the principal entry point of contamination which on exposure to heavy metals multiplied causing hyperplasia.

In the present study the liver of fish also exhibited the severe histopathological lesions like deshaped hepatocytes, vacuolization, necrosis of parenchyma, pyknosis and infiltration of leucocytes. Similar results have been reported in liver of different fishes, *Oreochromis mossambicus* exposed to cadmium and zinc (Van Dyk et al. 2007), *Clarias gariepinus* exposed to fuel oil for 14 days (Gabriel 2007),*Channa punctatus* exposed to hexavalent chromium (Mishra and Mohanty 2008). *Clarias batrachus* to $ZnSO_4$ (Prasanna Subhas Joshi 2011), *Cyprinus carpio* to lethal concentrations of Cr (Parvathi et al.2011), *Tilapia zilli* to Al (Hadi and Alwan 2012), *Clarias gariepinus* to sewage/ domestic wastewater containing Cu, Fe, Pb, Cd, Mn and Zn (Authman et al. 2013). It has been noticed that only toxicant exposed liver show vacuolation and pyknosis (Karyomegaly). Degeneration of liver tissue and necrosis could be due to the infiltration of leucocytes and according to Hughes et al. (1979) necrosis is the direct toxic effect of the pollutant. Further, the histopathological picture of the liver of *M. armatus* corroborates with the biochemical changes accounting for the functional disruption on the activity of the organ due to cellular damage.

Similarly, kidney of *M. armatus* exhibits dilation and vacuolation of kidney tubules, hypertrophy of renal tubules, degeneration of glomeruli and necrosis of hematopoietic tissue. These degenerative changes in kidney were also reported in various fishes such as *Anabas testudineus* exposed to unused lignite mine (Supap Saenphet 2009), *Clarias batrachus* to $ZnSO_4$ (Prasanna Subhash Joshi2011), *Cyprinus carpio* to lethal concentrations of Cr (Parvathi et al. 2011). The main function of kidney is washing/filtration of body fluids and to maintain the homeostasis. Severity of lesions observed in the present study showed that uriniferous tubules and hematopoietic tissue was badly damaged which could impair the renal function and as a consequence heavy metals get accumulated in various organs or muscle of the fish.

CONCLUSIONS

The effluent from the power plant containing heavy metals influenced the water quality of the rivulet under study. Concentrations of heavy metals assessed Fe, Ni, Mn and Cr were found to exceed the permissible limits set for water quality for ecosystem and human health. These constructions not only influence quality of water, but they directly influence flora and fauna where in this case it was observed that a major protein source in the form of fish *Mastacembelus armatus* was influenced which is the major diet of local population. These heavy metals also influence fish physiology. Hence, some scientific method of detoxification is essential to improve the health of these economic fish in the stressed environmental conditions. Since virtually all metals investigated were found in higher concentration, so government should intact laws which will ensure that industries make use of standard waste treatment plants for the treatment of their wastes before they are being discharged into water bodies. Some monitoring programs should also be launched from time to time in order to prevent the misuse of valuable water resources, to check their water quality status, and to restore them for the welfare of society and to protect the natural environment. Since these fishes share the local market therefore they must be screened by food agencies before they reached the humans in order to avoid the epidemics as occurred in Japan (1956) due to consumption of heavy metal contaminated fish and fishery products. Further studies are suggested particularly on the reproductive aspects of the fish in order to check its reproductive health/potential which will help to conserve the species.

AUTHORS' CONTRIBUTIONS

MJ carried out the whole experimental work and drafted the manuscript. NU guiding the protocol and assist during the sampling, experiments and also participated in finalizing the manuscript. Both authors read and approved the final manuscript.

ACKNOWLEDGEMENTS

The authors wish to acknowledge the Chairman Department of Zoology for making available all the necessary facilities. Chairman Department of Geology is acknowledged for permitting the use of Atomic Absorption Spectrophotometer in his department. Dr. Shakil Ahmad, incharge of Geochemistry Lab, Department of Geology, A.M.U., is also acknowledged for his cooperation during the analysis.

REFERENCES

1. Abdul Q, Riffat NM (2011) Heavy metals in eight edible fish species from two polluted tributaries (Aik and Palkhu) of the River Chenab, Pakistan. Biol Trace Elem Res 143:1524-1540
2. Agency for Toxic Substances and Disease Registry (ATSDR) (2004) Agency for toxic substances and disease registry. Clifton Road, Atlanta: Division of Toxicology. Available at http://www.atsdr.cdc.gov/toxprofiles
3. Agency for Toxic Substances and Disease Registry (ATSDR) (2005) Agency for toxic substances and disease registry. Clifton Road, Atlanta: Division of Toxicology. Available at http://www.atsdr.cdc.gov/toxprofiles
4. American Public Health Association (APHA) (2005) Standard methods for the examination of water and wastewater analysis, (21st ed.). Washington D.C: American Water Works Association/ Water Environment Federation. p 289
5. Anim EK, Ahialey EK, Duodu GO, Ackah M, Bentil NO (2010) Accumulation profile of heavy metals in fish samples from Nsawam, along the Densu River, Ghana. Res J Environ Earth Sci 1:56-60
6. Authman MMN, Ibrahim SA, El-Kasheif MA, Gaber HS (2013) Heavy metals pollution and their effects on gills and liver of the Nile catfish (*Clarias gariepinus*) inhabiting El-Rahawy Drain, Egypt. Global Veter 10:103-115

7. Baki AS, Dkhil MA, Al-Quraishy S (2011) Bioaccumulation of some heavy metals in tilapia fish relevant to their concentration in water and sediment of Wadi Hanifah, Saudi Arabia. African J Biotech 10:2541-2547
8. Boeck GD, Meeus W, Coen WD, Blust R (2004) Tissue specific Cu bioaccumulation patterns and differences in sensitivity to water borne Cu in three fresh water fish: rainbow trout (*Oncorhyncus mykiss),* common carp (*Cyprinus carpio),* and gibel carp (*Carassius auratus*). Aquat Toxicol 70:179-188
9. Carrol NV, Longley RW, Roe JH (1956) Glycogen determination in liver and muscle by use of anthrone reagent. J Biol Chem 22:583-593
10. Coetzee L, du Preez HH, Van Vuren JHJ (2002) Metal concentrations in *Clarias gariepinus* and *Labeo umbratus* from the Olifants and Klein Olifants river, Mpumalanga, South Africa: zinc, copper, manganese, lead, chromium, nickel, aluminium and iron. Water SA 28:433-448
11. Duncan DB (1955) Multiple ranges and multiple F- tests. Biometrics Fisheries Resources 11:1-42
12. Eisler R (1988) Zinc hazards to fish, wildlife and invertebrates: A synoptic review. US Fish Wildlf Survev Reprod Biol 85:1-6
13. Emere MC, Dibal DM (2013) Metal accumulation in some tissues/organs of a freshwater fish (*Clarias Gariepinus*) from some polluted zones of River Kaduna. J Biol Agri Healthcar 3:112-117
14. FAO (Food and Agriculture Organization) (1983) Compilation of legal limits for hazardous substances in fish and fishery products. FAO Fish Circular 464:5-100
15. FAO/WHO (1989) National Research Council Recommended Dietary Allowances (10th ed). Washington, DC. USA: National Academy Press.
16. Fatima M, Usmani N (2013) Histopathology and bioaccumulation of heavy metals (Cr, Ni and Pb) in fish (*Channa striatus* and *Heteropneustes fossilis*) tissue: A study for toxicity and ecological impacts. Pakistan J biol Sci 16:412-420
17. Gabriel UUU, Ezeri GNO, Amakiri EU (2007) Liver and kidney histopathology: Biomarkers of No. 1 fuel toxicosis inAfrican cat fish *Clarias gariepinus*. J Anim Veter Advan 6:379-384

18. Hadi AA, Alwan SF (2012) Histopathological changes in gills, liver and kidney of fresh water fish, *Tilapia zilli,* exposed to aluminum. Int J Pharma Life Sci 3:2071-2081

19. Hughes GM, Perry SF, Brown VM (1979) A morphometric study of the effects of nickel, chromium and cadmium on the secondary lamellae of rainbow trout gills. Water Res 13:665-679

20. Humason GL (1979) Animal tissue techniques. U.S.A: W.H. Freeman and company. p 661

21. Jabeen F, Chaudhry AS (2010) Monotoring trace metals in different tissues of *Cyprinus carpio* from Indus River in Pakistan. Environ Monit Assess 170:645-656

22. James R, Sampath K, Alagurathinam S (1996) Effects of lead on respiratory enzyme activity, glycogen and blood sugar levels of the teleost *Oreochromis mossambicus* (Peters) during accumulation and depuration. Asian Fishery Sci 9:87-100

23. Javed M (2012) Tissue-specific bio-accumulation of metals in fish during chronic waterborne and dietary exposures. Pakistan Veter J 32:567-570

24. Javed M, Usmani N (2011) Accumulation of heavy metals in fishes: a human health concern. Int J Environ Sci 2:659-670

25. Javed M, Usmani N (2012) Uptake of Heavy Metals by Channa Punctatus from Sewage–Fed Aquaculture Pond of Panethi, Aligarh. Global J Res Engg (C) 12:27-34

26. Javed M, Usmani N (2012) Toxic effects of heavy metals (Cu, Ni, Fe, Co, Mn, Cr, Zn) to the haematology of *Mastacembelus armatus* thriving in Harduaganj Reservoir, Aligarh, India. Global J Medical Res 12:59-64

27. Javed M, Usmani N (2013) Haematological indices of *Channa punctatus* as an indicator of heavy metal pollution in waste water aquaculture pond, Panethi, India. African J Biotech 12:520-525

28. Javed M, Usmani N (2013b) Investigation on accumulation of toxicants and health status of freshwater fish *Channa punctatus,* exposed to sugar mill effluent. Int J Zoo Res 3:43-48

29. Joshi PS (2011) Studies on the effects of zinc sulphate toxicity on the detoxifying organs of fresh water fish Clarias batrachus (Linn.). Gold Res Thoughts 1:1-4

30. Madhusudan S, Fatma L, Nadim C (2003) Bioaccumulation of Zn and Cd in fresh water fishes. Indian J Fish 50:53-65

31. Maruthi YA, Subba Rao MV (2000) Effect of distillery effluent on biochemical parameters of fish, Channa punctatus (Bloch). J Environ Pollut 7:111-113

32. Mishra A, Mohanty B (2008) Acute toxicity impacts of hexavalent chromium on behavior and histopathology of gill, kidney and liver of the freshwater fish, *Channa punctatus* (Bloch). Environ Toxicol Pharmacol 26:36141

33. Murugan SS, Karuppasamy R, Poongodi K, Puvaneswari S (2008) Bioaccumulation Pattern of Zinc in Freshwater Fish *Channa punctatus* (Bloch.) After Chronic Exposure. Turkish J Fish Aquat Sci 8:55-59

34. Nath K, Kumar N (1988) Hyperglycemic response of *Heteropneustes fossilis* exposed to nickel. Acta Hydro et Hydrobiol 16:333-336

35. Nwajei GE, Obi–Iyeke GE, Okwagi P (2012) Distribution of Selected Trace Metal in Fish Parts from the River Nigeria. Res J Recent Sci 1:81-84

36. Osman Alaa GM, Werner K (2009) Water quality and heavy metal monitoring in water, sediments, and tissues of the African Catfish *Clarias gariepinus* (Burchell 1822) from the River Nile, Egypt. J Environ Protect 1:389-400

37. Palanisamy PG, Sasikala D, Mallikaraj NB, Natarajan GM (2011) Electroplating industrial effluent chromium induced changes in carbohydrates metabolism in air breathing cat fish *Mystus cavasius*(Ham). Asian J Exp Biol Sci 2:521-524

38. Parvathi PK, Mathan S, Sarasu R (2011) Sublethal effects of chromium on some biochemical profiles of the freshwater teleost, *Cyprinus carpio*. Int J App Biol Pharma Techno 2:295-300

39. Pickering QH (1974) Chronic toxicity of nickel to the fathead minnow. J. WPCF 46:660-765

40. Radhakrishaniah K, Venkataramana P, et al. (1992) Effect of lethal and sublethal concentrations of copper on glycolysis in the liver of the fresh water teleost, Labeo rohita (Ham). J Environ Biol 13:63-68

41. Ravanaiah G, Narasimha Murthy CV (2010) Impact of aquaculture and industrial pollutants of Nellore district on the histopathological changes in the gill of fish, *Tilapia mossambica*. Ind J Comp Animal Phsiol 28:108-114

42. Robinson J, Avenant- Oldewage A (1997) Chromium, Copper, Iron and Manganese bioaccumulation in some organs and tissues of *Oreochromis mossambicus* from the lower Olifants River, inside the Kruger National Park. Water S A 4:387

43. Selda TO, Zan E, Ismail K (2005) Comparative study on the accumulation of heavy metals in different organs of tench (*Tinca tinca* L.1758) and plerocercoids of its endoparasite *Ligula intestinalis*. Parasitol Res 97:1412-1419

44. Sesha Srinivas V, Rao BM (1998) Chromium induced alterations in the gill of the freshwater teleost fish, *Labeo rohita*. Ind J Compar Anim Physiol 17:31-33

45. Smith IC, Carson BL (1981) Trace metals in the Environment, Cobalt. Ann Arbor, MI: Ann Arbor Science Publishers, Inc. (6), 1202

46. Srivastava R, Srivastava N (2008) Changes in nutritive value of fish, *Channa punctatus* after chronic exposure to zinc. J Environ Biol 29:299-330

47. Stokes PM (1979) Copper accumulations in aquatic biota. In: JO Nriagu (ed.) Copper in the environment. Part 1 Ecological cycling. New York, United States of America: Wiley. pp 358-381

48. Supap Saenphet, Thaworn W, Saenphet K (2009) Histopathological alterations of the gills, liver and kidneys in *Anabas testudineus* (Bloch) fish living in an unused lignite mine LI district. Lamphun Province, Thailand SouthEast Asian J Trop Med Public Health 40:1121-1126

49. Taweel AKA, Shuhaimi-Othman M, Ahmad AK (2012) Analysis of heavy metal concentrations in tilapia fish (*Oreochromis niloticus*) from four selected markets in Selangor, Peninsular Malaysia. J biol Sci 12:138-145

50. UNEPGEMS (2006) United Nations Environment Programme Global Environment Monitoring System/ Water Programme (Water Quality for Ecosystem and Human Health).

51. Usero J, Gonza´lez-Regalado E, Gracia I (1997) Trace metals in the bivalve molluscs Ruditapes decussatus and Ruditapes philippinarum from the Atlantic Coast of Southern Spain. Environ Int 23:291-298
52. USFDA (1993b) Food and drug administration, Guidance document for nickel in shell fish. Washington D.C: DHHS/PHS/FDA/CFSAN/office of seafood.
53. Van Dyk JC, Pieterse GM, Van Vuren JHJ (2007) Histological changes in the liver of *Oreochromis mossambicus* after exposure to cadmium and zinc. Ecotox Environ Saf 66:432-440
54. Van Vuren J, Van der Merwe M, Du Preez H (1994) The effect of copper on the blood chemistry of*Clarias gariepinus* (clariidae). Ecotoxicol Environ Saf 29:187-199
55. Voynar AI (1960) Biologicheskaya Rol'mikroelementov v Organizme Zhivotnykh I Cheloveka Biological Function of Trace elements in Animals and Man). Moscow: Vyssh, shk. Press.
56. Yousafzai AM, Douglas P, Chivers KAR, Ahmad I, Siraj M (2010) Comparison of heavy metals burden in two freshwater fishes *Wallago attu* and *Labeo dyocheilus* with regard to their feeding habits in natural ecosystem. Pakistan J Zoo 42:537-544

Citations

CHAPTER 1

Eric B Alsop, Eric S Boyd, and Jason Raymond, Merging Metagenomics and Geochemistry Reveals Environmental Controls on Biological Diversity and Evolution, doi: 10.1186/1472-6785-14-16.

CHAPTER 2

Isinkaye M Omoniyi, Shitta M B Oludare, and Oderinde M Oluwaseyi, Determination of Radionuclides and Elemental Composition of Clay Soils by Gamma- and X-Ray Spectrometry, doi:10.1186/2193-1801-2-74.

CHAPTER 3

Stuart J Self, Bale V Reddy, and Marc A Rosen, Review of Underground Coal Gasification Technologies and Carbon Capture, doi: 10.1186/2251-6832-3-16.

CHAPTER 4

Andrea Baglieri, Daniela Vindrola, Mara Gennari, and Michèle Negre, Chemical and Spectroscopic Characterization of Insoluble and Soluble Humic Acid Fractions at Different pH Values, doi:10.1186/s40538-014-0009-x.

CHAPTER 5

Giuseppe Sappa, Sibel Ergul, and Flavia Ferranti, Geochemical Modeling and Multivariate Statistical Evaluation of Trace Elements in Arsenic Contaminated Groundwater Systems of Viterbo Area, (Central Italy), doi:10.1186/2193-1801-3-237.

CHAPTER 6

Paula B Miqueletto, Fernando D Andreote, Armando CF Dias, Justo C Ferreira,EugênioJ V dos Santos Neto, and Valéria M de Oliveira, Cultivation-Independent Methods Applied to the Microbial Prospection of Oil and Gas in Soil from a Sedimentary Basin in Brazil, doi:10.1186/2191-0855-1-35.

CHAPTER 7

Ibrahima Youm, Xiaoyan Yang, James B Murowchick and Bi-Botti C Youan, Encapsulation of docetaxel in oily core polyester nanocapsules intended for breast cancer therapy, doi:10.1186/1556-276X-6-630.

CHAPTER 8

Mehjbeen Javed and Nazura Usmani, Assessment of Heavy Metal (Cu, Ni, Fe, Co, Mn, Cr, Zn) Pollution in Effluent Dominated Rivulet Water and Their Effect on Glycogen Metabolism and Histology of Mastacembelus Armatus, doi:10.1186/2193-1801-2-390.

Index